INTERNATIONAL ENERGY AGENCY

ELECTRICITY IN INDIA

Providing Power for the Millions

INTERNATIONAL
ENERGY AGENCY
9, rue de la Fédération,
75739 Paris, cedex 15, France

ORGANISATION FOR
ECONOMIC CO-OPERATION
AND DEVELOPMENT

The International Energy Agency (IEA) is an autonomous body which was established in November 1974 within the framework of the Organisation for Economic Co-operation and Development (OECD) to implement an international energy programme.

It carries out a comprehensive programme of energy co-operation among twenty-six* of the OECD's thirty Member countries. The basic aims of the IEA are:

- To maintain and improve systems for coping with oil supply disruptions;
- To promote rational energy policies in a global context through co-operative relations with non-member countries, industry and international organisations;
- To operate a permanent information system on the international oil market;
- To improve the world's energy supply and demand structure by developing alternative energy sources and increasing the efficiency of energy use;
- To assist in the integration of environmental and energy policies.

*IEA Member countries: Australia, Austria, Belgium, Canada, the Czech Republic, Denmark, Finland, France, Germany, Greece, Hungary, Ireland, Italy, Japan, the Republic of Korea, Luxembourg, the Netherlands, New Zealand, Norway, Portugal, Spain, Sweden, Switzerland, Turkey, the United Kingdom, the United States. The European Commission also takes part in the work of the IEA.

Pursuant to Article I of the Convention signed in Paris on 14th December 1960, and which came into force on 30th September 1961, the Organisation for Economic Co-operation and Development (OECD) shall promote policies designed:

- To achieve the highest sustainable economic growth and employment and a rising standard of living in Member countries, while maintaining financial stability, and thus to contribute to the development of the world economy;
- To contribute to sound economic expansion in Member as well as non-member countries in the process of economic development; and
- To contribute to the expansion of world trade on a multilateral, non-discriminatory basis in ac-cordance with international obligations.

The original Member countries of the OECD are Austria, Belgium, Canada, Denmark, France, Germany, Greece, Iceland, Ireland, Italy, Luxembourg, the Netherlands, Norway, Portugal, Spain, Sweden, Switzerland, Turkey, the United Kingdom and the United States. The following countries became Members subsequently through accession at the dates indicated hereafter: Japan (28th April 1964), Finland (28th January 1969), Australia (7th June 1971), New Zealand (29th May 1973), Mexico (18th May 1994), the Czech Republic (21st December 1995), Hungary (7th May 1996), Poland (22nd November 1996), the Republic of Korea (12th December 1996) and Slovakia (28th September 2000). The Commission of the European Communities takes part in the work of the OECD (Article 13 of the OECD Convention).

FOREWORD

The IEA and India have conducted a mutually beneficial dialogue on energy policy for the past five years. Our exchanges were reinforced in number and depth after the signing of a Declaration of Co-operation in the Field of Energy in May 1998.

Because of India's growing external energy dependency and global emissions from the energy sector, the changes taking place there are increasingly important to world energy markets. India is the third-largest producer of hard coal after China and the United States. India imports around 1.4 million barrels of oil per day, 60 per cent of its total needs. This dependency is projected to grow to 85 per cent by 2010 and to over 90 per cent by 2020. India's crude oil imports are projected to reach 5 million barrels per day in 2020, which is more than 60 per cent of current Saudi Arabian oil production. Energy and electricity will be required for a population that exceeded one billion in 2000 and to fuel an economy that grew at an average annual rate of 7 per cent from 1993 to 1997.

India's central and state governments have begun efforts to reform the power sector. The liberalisation of India's electricity market, initiated a decade ago, was expected to rationalise consumption and improve the allocation of financial and energy resources. But the Indian power sector is now facing a serious crisis with implications for the country's overall economic growth and development.

The IEA and its Member countries stand ready to assist the Government of India in its efforts to improve the overall functioning of the electricity sector. This study was undertaken to provide an assessment of India's electricity liberalisation efforts. It is similar to the energy policy surveys carried out regularly by the IEA in its Member countries and includes recommendations. The study aims to contribute to the debate on electricity liberalisation policy and to the Government of India's efforts to develop and improve the efficiency of the Indian electricity market.

The reader may well note a difference between the analysis in this book and that in other IEA publications, in that we here recommend active government intervention to help prepare the way for an India-wide free power market. The Agency holds firmly to the view that markets work best when they respond to market forces alone, and this will be the case even for India in the future. For the moment, however, no free market exists in India. Only the central government can create the preconditions for a market to emerge, by inducing the states to reform their bankrupt utilities, by encouraging trade among states, by promoting an environment that will attract investment from home and abroad. Once these tasks are achieved, we look to private investment and private enterprise to provide India's billion citizens with ample electricity at affordable prices.

Robert Priddle
Executive Director

ACKNOWLEDGEMENTS

This study was produced by the International Energy Agency's Office of Non-Member Countries. The principal authors of this book are Pierre Audinet and François Verneyre.

The IEA wishes to acknowledge the co-operation of the Government of India, and in particular the Ministry of Power, for its help in arranging meetings and providing information during the IEA team's visit to New Delhi in November 2000, for its support in documenting IEA analysis and for providing comments on the draft.

Reviewers of the manuscript included: Olivier Appert (IEA), Norio Ehara (IEA), Ralf Dickel (IEA), Carlos Ocana (IEA), Jean-Pierre Sérusclat (EDF), Richard Perrier (EDF), and several others.

Special thanks are given to the following IEA contractors: Trevor Morgan for an update on electricity subsidies and Anouk Honoré for her assistance in gathering information. Chantal Boutry warrants special mention for her secretarial help.

TABLE OF CONTENTS

ANNEXES

List of Tables

List of Figures

List of Boxes

List of Maps

EXECUTIVE SUMMARY AND POLICY MESSAGES

India's electricity-supply industry is mainly owned and operated by the public sector. It is currently running a growing risk of bankruptcy. This has created a serious impediment to investments in the sector at a time when India desperately needs them. This is reflected in the sharp decrease of the ratio of electricity consumption growth to GDP growth in the 1990s. In other words, in the past decade, electricity consumption growth did not follow economic growth. For 1991-1999, the elasticity of electricity consumption with regard to GDP was 0.97 when it was 2.1 for Korea and 0.99 for the OECD on an average.[1] Neither the high structural needs of the Indian economy, nor improvements in energy efficiency can explain this low figure. It is a reflection of an increasing gap between supply and demand, the continuously deteriorating quality of power[2], and a low level of access to electricity. It is also the result of large investments made by the manufacturing sector in stand-by and stand-alone facilities to compensate these deficiencies. Unless strong measures are taken immediately to correct this trend, India's overall economic development will be slowed.

The central issue now is how to enable power utilities to earn a return on investment. Price levels are too low for the system to be financially viable. In the Indian states, vested political interests impede utilities from collecting revenue. They maintain a price structure with large and unjustifiable subsidies. Politicians often interfere in the management of power utilities, hindering their efforts to curb power theft. As a result, transmission and distribution losses in India have increased, further eroding the financial situation of the state electricity utilities. These are not new trends, but the situation has reached a critical stage, where the government can no longer cover the losses of the state power utilities. For decades, the costs incurred for the development and operation of the electrical system increased faster than general economic growth, outstripping public finances' ability to make up for uncollected rates. The central government has for a long time given priority to developing access to electricity. At the state level this meant low prices for domestic and agriculture consumers and relatively higher prices for electricity supplied to the industry and commercial sectors. Even this system did not compensate for the subsidy burden. The government was obliged to compensate the difference. Growth of the electricity sector has outstripped the growth of the public money available to bear the cost of the increasing subsidies; the mechanism is not sustainable. Nonetheless, consumers who are used to low prices, and populist politicians resist change.

---

1. For 1971-1990, the figures were respectively of 1.7, 1.6 and 1.1 (Source: IEA).
2. With high voltage fluctuations and recurring black-outs.

To face increasing investment needs, the central government began in 1991 to focus on attracting private investment. The aim was to sustain power-sector development while keeping public expenditure under control. Competition was gradually introduced in bidding for generation projects. Since the mid-1990s, in response to the growing financial difficulties of state electricity boards (SEBs), the World Bank recommended introducing private capital into the power distribution sector and a new regulatory framework, which would allow independent tariff-setting to correct large price distortions. The central government established the legal framework for this new arrangement in 1998. More recently, it has focussed on distribution, trying to increase revenue collection and additional capital.

Unfortunately, the results of this decade of reform have fallen well below expectations and the central government now seems short of solutions. It is probably too early to judge the final outcome of the change, from a command-and-control public-dominated model to a more market-determined sector. However, the present indicators point to the need for urgent action. In 1995-1996, nine of the 19 SEBs incurred losses. In 2000-2001, all of them were in the red. SEBs are increasingly unable to pay for the electricity they purchase from the central public-sector power companies[1], or from independent power producers (IPPs). The official – and probably underestimated – figure for transmission and distribution (T&D) losses is higher than ever, reaching 25% in 1997-98. In such conditions, the much-expected private investment has been well below expectations, and even public investments were relatively lower in recent years than before. The difficulties experienced by several private investors[2] have discouraged potential additional investors. Unless radical measures are taken in the very short term, there is a real risk of stagnation in investment in the whole system. The demand-supply gap will continue to grow. With negligible new private investment in generation or distribution, and the central and state governments' shrinking ability to develop and maintain the power system, an increasing number of consumers will be driven to invest in stand-by or stand-alone generation sets at the expense of the public interest, challenging the very roots of social and regional equity.

Under the Constitution, electricity is on the "concurrent list", which means that the states, rather than the central government, are primarily responsible for setting electricity tariffs. The states have the largest share of generation and transmission assets and almost all distribution in their control. The states have a key role to play in effecting institutional and result-oriented changes. However, the IEA believes the role of the central government is vital in guiding the developments to come and especially in providing the necessary legal and financial incentives for the states to implement reforms. The following policy recommendations are addressed to the central government. The IEA believes the past reform policies overlooked the political and technical obstacles to overhauling the existing system, and the time needed to do so. The case of Orissa clearly demonstrates that privatisation cannot in itself sustain the sector's development. Competition and private investment alone cannot be expected to resolve management issues, market distortions and the interference of vested political interest in the system.

---

1. Together, they owed the central public-sector electricity companies more than seven billion dollars as of March 2001 (GOI, 2001a).
2. Such as AES or BSES in distribution in Orissa, and Enron in generation in Maharashtra.

The existing public electricity-supply industry needs to be put in order first to allow the private sector to operate. To ensure an optimal allocation of capital and energy resources, the size of electricity markets at the state level is still too small. Efforts must be made to improve the development and the management of the power sector at the Union level. The central government has underestimated the specific regulatory needs for competition to expand and for the grid to develop in a sustainable manner.

In priority order, the central government should concentrate on each of the following tasks for the next five years:

■ **Adopt a comprehensive reform plan for the electricity supply industry** introducing competition to the electricity sector and improving its overall performance, taking into account the goals of electricity access, energy security, environmental protection and economic growth.

■ **Make the states accountable for the performance of their public electricity system,** by providing additional financial incentives to better-performing states on the basis of a transparent set of criteria. Absolute priority should be given to achieving full cost recovery within a defined time frame. This is necessary because so-called non-technical losses – actually unpaid or stolen electricity – are largely the result of political interference or of negligence by the state governments. State governments should be provided incentives to enforce the law and to clamp down on non-paying consumers – and that they be punished if they do not. States could be obliged to account for the cost of transmission and distribution (T&D) losses in their budgets and to establish tariffs based on a low level of T&D losses (including theft). This would reinforce the responsibilities of regulators to monitor T&D losses and differentiate between technical and non-technical losses. A legal framework should be established to sanction the loose handling of tariffs and power theft, and providing targets and incentives for the states. No progress can be achieved without improved revenue collection from final consumers.

■ **Set and adhere to a firm timetable for introducing market mechanisms.** An implementation timetable should provide for establishing the regulatory framework, reforming subsidies, curbing power theft and developing innovative solutions for private-sector distribution. The focus should be on development of a power market at the central (Union) level and should clearly identify the steps to be taken at the state-level. A number of such mechanisms have nominally been implemented, but future implementation will need clearly-designed monitoring criteria.

■ **Concentrate political accountability in a single energy ministry.** Integration of political accountability into a single energy ministry is essential. Only an integrated authority can exploit economies of scale through co-operation and integration at the Union level.

■ **Facilitate the mobilisation of investment-capital by the centralised public utilities.** The strategy of increasing generation capacity through large-scale IPPs has not proved successful. The gap has been partially covered by the development of self-generation and by central public-sector investments. It would be beneficial to have a mix of large and small public-utility capacity, at least temporarily, to reduce the supply-demand gap.

■ **Facilitate and encourage grid access for surplus electricity from auto-producers (captive producers), while encouraging private investment in generation.** Use of existing auto-production capacity should be maximised. This new capacity could make a significant contribution from the private sector to the growing electricity market.

■ **Create the framework for a power market at the Union level.** This recommendation complements the above call for more integration. India is still far from the point where a competitive market can govern the supply-demand balance. However, India's eventual target should be an electricity market. Only at the aggregated Union level are there sufficient demand and supply. The first steps toward such a national market would be increased investment in the Union-level electricity grid, and giving more freedom for market players to exchange and trade across state borders.

■ **Implement measures to improve business practices of the electricity supply industry in the public sector. More attention should be paid to developing human skills and personal accountability** at all levels of the state and central government.

Map 1 States of India

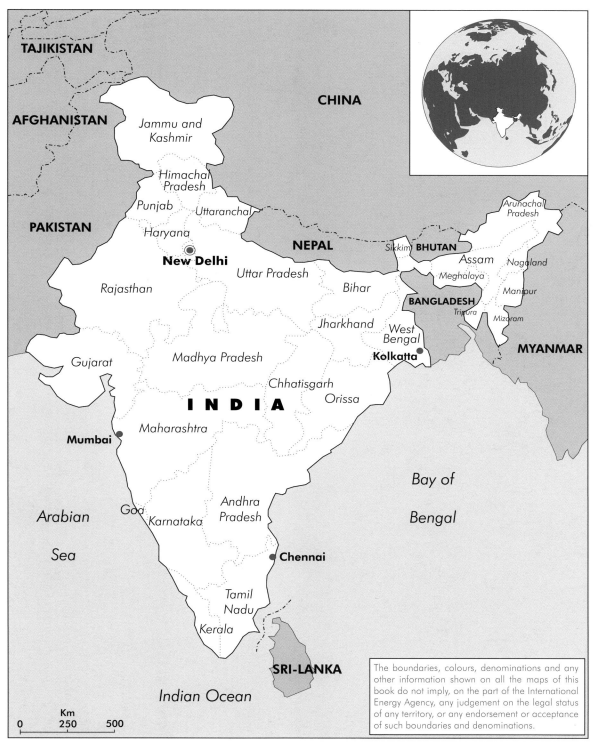

Source: IEA.

I. CONCLUSIONS AND POLICY RECOMMENDATIONS

This review focuses on the electricity-supply industry. It explores the linkages between price, demand and supply. It recounts the emergence of a competitive power market so far and examines possible future development of this market.

The recommendations that follow are based on the tenets of the IEA "Shared Goals" adopted in 1993. These goals are loosely used in this review as a set of criteria to analyse the Indian electricity sector. India faces many challenges on the energy front, but must deal simultaneously with economic, social and environmental challenges as well. These considerations weigh on energy policy formulation. Wider access to modern commercial energy sources, the reduction of airborne pollutants in cities, and improvements in the reliability and the quality of energy services are strong drivers of Indian energy policy.

In addition to general policy issues, we review India's electricity-supply industry and its various market components or sub-sectors, examining the demand side of the power market, then distribution, transmission and generation. Rural electrification is considered in its own rights, as it involves specific market dynamics.

This book does not promote a pre-determined model of market and regulatory organisation. It does however try to identify possible areas of improvement of the existing set-up, in order to restore the sector's health, introduce more competition and satisfy the country's growing electricity needs.

GENERAL ELECTRICITY POLICY

Issues

India ranked eighth in the world in total electricity generated in 1998 – between France and the United Kingdom – with about 494 TWh. But because of India's large population, consumption of electricity per capita was only 460 kWh/year – among the lowest in the world. The world average is 2,252 kWh per capita.

The IEA's *World Energy Outlook 2000* (IEA, 2000a) projects an average annual growth rate of 4.9% for the next 20 years for India's GDP and of 5.2% for electricity generation, corresponding to a threefold rise in electricity supply over the period. These projections suppose high levels of investment throughout the entire sector.

Indian electricity policy aims to provide cost-effective, affordable and secure access to electricity for all. Given the large rural population, rural electrification remains an important objective. The Government of India, in conjunction with state governments, should:

■ use indicative planning and the normal techno-economic project clearances to guide technical choices and the power mix;

■ define and implement the principles on which tariffs are set and tariffs themselves, and do so increasingly through independent regulatory commissions;

■ increase the level of public funds in the sector to encourage development;

■ create an environment that will attract private players.

Most of the electricity-supply industry in India remains in the public sector. The central government, through public companies, owns and operates one-third of the power generation and interstate exchanges. At the state level, SEBs own and operate most of the remaining two-thirds of the generation capacity, as well as single-state transmission and distribution systems. States define their own tariff structures.

State-wide electricity systems are relatively small, in line with the low country per-capita electricity consumption. The sector would benefit from further national integration and the economies of scale that would accompany it.

Although the central government is politically committed to reforming the regulatory framework to facilitate the development of a power market, implementation of the reform policies has been slow.

Power-sector policies are designed and implemented by the Ministry of Power at the national level and by ministries in charge of power or energy at the state level. Political intervention in electricity matters is common at the national level. Fuel-supply issues for power generation projects may also need clearance from the Ministry of Coal (and the Ministry of Railways) or the Ministry of Petroleum and Natural Gas, whose opinions often differ from those of the Ministry of Power. Political intervention motivated by social concerns is frequently exerted and can frustrate efforts to rationalise electricity prices.

The SEBs are effectively bankrupt as a result of political interference and mismanagement. For the last several years, their revenues have been insufficient to cover the costs of providing electricity. Arrears owed to the central generating companies are now equivalent to the cost of one year of consumption. This is a crucial issue, but not the only one. A second problem is the ineffective decision-making process in the SEBs. There is no incentive for technical staff to implement or run cost-effective operations. The results of these two failures are unsatisfied demand, poor quality of electricity and unreliable supply. The SEBs' inefficiencies and ineffectiveness clearly impair the sustainable development of the Indian power sector and act as a bottleneck for economic growth and development.

Policy changes to reform India's electricity-supply industry have been initiated for some time[1]. They were developed with the support of multilateral agencies such as the World Bank. Since 1991, the government has promoted private-sector participation in the generation sector as a cost-effective means to build-up additional capacity. Incentives are planned to encourage SEBs to improve the efficiency of existing generating capacity. The new policies favour the unbundling of SEBs and the privatisation of distribution. New regulatory institutions are also being established at the national and state levels: the electricity regulatory commissions (ERC). The most recent debate concerns the "Electricity Bill", which would further pave the way to a competitive power market in India. But, despite these good intentions and a number of valuable legal and institutional changes, the implementation of corresponding policy measures has been slow.

Public opinion in India appears increasingly aware of the adverse effects of populist measures resulting in the underpricing of electricity and degraded service. There is a realisation of the need for an economically-viable power sector. This gradual change of opinion should put pressure on political leaders.

Recommendations

■ In its pursuance of sustainable energy development, the Government of India should identify, define and separate economic, social and environmental policy objectives.

■ There is a crucial need to reduce subsidies by central and state governments to the electricity sector. Accelerating subsidy reductions will decrease the burden on public finances freeing money for public investment. Given the paucity of public finances in the states, subsidising electricity consumption is unsustainable. Where subsidies exist, they should be of limited amounts and duration. Market players (SEBs in particular) should not have to bear their cost. Cross-subsidies should be avoided.

■ The electric power system in this emerging economy requires considerable investment. The central government should promote market expansion to facilitate investment, rationalise tariffs, and develop competition.

■ Specifically, the central government should foster the technical, economic and institutional integration of the electricity-supply industry. Consolidated political leadership is needed for the whole energy sector. This leadership would bring together the often disparate and competing offices responsible for the development of electricity, coal, and water resources for hydroelectricity.

■ The Indian transmission system must be expanded to reduce shortages, facilitate competition and respond to growth in demand. This requires the development of a national network of transmission lines (400 kV and above), enabling numerous power exchanges between states.

---

1. Opening of generation to private-sector participation in 1991; Common Minimum National Action Plan for Power in 1996; Resolutions of the Conference of Chief Ministers in 1998; and legal innovations introduced by the Electricity Regulatory Commissions Act of 1998.

■ The SEBs need to have an entrepreneurial focus and should be corporatised to improve their efficiency and financial leverage. Sound internal business practices and specific cost reduction goals should be introduced to guide the decision-making process. Within the SEBs or their corporatised entities, the various hierarchical levels should be better managed and given more financial autonomy. This is particularly true for the distribution sector, where a great number of individual decisions has to be made. The decision-making process should be streamlined, and staff should be rewarded for improvements in cost efficiency.

■ Principles of accountability and independence from political interference should be enforced in SEBs. The functions of generation, transmission and distribution should be separated vertically in all Indian states so that cost components and profit centers can be identified. Separate companies could be created to fulfill each function.

■ The central government should control the effective implementation of the existing plans for reforming the electricity industry, including using financial rewards or penalties. Clear deadlines and strict monitoring of states' actions are required. New plans and measures should also be well-monitored.

■ The government needs to ensure improvement in data quality, availability and transparency. Currently-available electricity data are not accurate enough. This is of special concern, because reporting at state level often underestimates losses and hampers the assessment of subsidies.

■ The government should design an integrated energy resource plan leading to a cost-efficient national power system. This is indispensable because of the existing market deficiencies in primary energy supply and the need to reduce the risk associated with the development of IPPs.

■ The private sector should be encouraged to participate at all levels of the supply chain (generation, transmission and distribution). Investment needs are too large to be met by public finances alone, and more competition is needed. An investment framework providing a secure, level playing field for private investors must be enforced. The process of contract allocation must be made transparent, and contractual terms must be enforced. The high risk perceived by foreign and domestic companies entering an emerging market needs to be better understood and analysed by the authorities so that risks are mitigated and fears allayed. This calls for stringent selection, prior clearance and improved definition of the terms of reference for projects before participation is invited.

RETAIL PRICING POLICY AND DEMAND

Issues

Most of the problems of the Indian power sector arise from the present retail pricing system and from the fact that too little of it is actually paid for. Out of total electricity generated, only 55% is billed and 41% is regularly paid for (GOI, 2001). Electricity is either stolen, not billed, or electricity bills are not paid. All this amounts to a mass of implicit subsidy. The financial burden thus created undermines the economic efficiency and viability of the electricity supply chain and is not in the long-term interests of consumers.

Retail tariffs in India (as well as bulk tariffs) are based on a cost-plus mechanism established at the time of India's independence in 1948. Electricity prices are subsidised for domestic consumers and for farmers.

Current retail prices of electricity represent less than 75% of real average costs. There is also a large amount of cross-subsidisation between consumer categories. The agriculture and household sectors are cross-subsidised by above-cost tariffs for commercial and industrial customers and railways[1]. The situation worsened in the 1990s. Official data demonstrate that subsidies to households trebled to 80.8 billion rupees over the period 1992-1993 to 1999-2000. Subsidies to agriculture more than tripled to 227 billion rupees over the same period. The government sought to justify these subsidies on social grounds but it clearly failed to achieve its social goal, as higher-income groups in fact appropriate most of the benefits since the subsidy is applied to the price of electricity within a given consumer-category, indifferently to the individual level of income.

Policies to achieve market pricing have been introduced in India. Central and state electricity regulatory commissions are slowly being established. They will issue tariff orders, and should eventually implement them. Policies to implement a minimum price have been pursued since 1996. The goal is a minimum price of 50 paise/kWh (1.1 US cents/kWh) for agriculture. But, delays in implementing such reforms have prevented even this simple goal from being met. Inflation (8% in 2000) has outpaced the growth in price per kilowatt hour. Policies in place also call for all end-use sectors ultimately to be charged at least 50% of the average cost of supply. This target was intended to be met in three years, but it has not yet been achieved in any of the states.

The poor cost-recovery rate, the very low price and the widespread non-payment of electricity are all deterrents to private investors. Investors cannot be assured that their applications for tariff increases to recover costs will be met, even by theoretically independent regulatory commissions.

The side effects of this way of subsidising energy consumption are significant. Overpricing of industrial electricity hampers competitiveness. In other sectors, underpricing of electricity is a direct incentive to waste power. Underpricing of electricity

---

1. In theory, cost-reflective tariff structures do not differentiate between final uses of electricity. The lowest tariffs apply to customers with the highest consumption and load factors (industrial customers). Households, on the contrary, pay the highest rate due to their low load factor, limited consumption and the relatively higher cost of distribution.

for agricultural use puts a heavy financial burden on the electricity sector and incidentally threatens water resources in the long-term. Low cost recovery translates into degradation of service, which in turn requires costly investments in stand-by capacity in the industrial and commercial sectors and by large domestic consumers. Confronted by high prices and unreliable supply from the network, big industrial consumers increasingly turn to "captive-power", which now represents more than one-third of their consumption. Every industrial consumer lost by the SEBs further worsens their financial situation since it reduces their sales base to low-paying customers. Subsidies artificially sustain demand from the consumers already connected to the grid, but a large unmet demand exists because of grid deficiencies and the inability of insolvent utilities to invest in additional connections.

Recommendations

■ State governments should promote and foster payment for electricity by all customers.

■ Legal action must be taken at the state level to prevent theft so that electricity suppliers have increased assurance that all customers will pay.

■ Tariff should be designed to recover costs on the basis of the electricity which is sold and paid for only, separating the cost of stolen electricity from the tariff structure. Otherwise, paying customers could end up being burdened with the costs of non-paying customers. To avoid such a difficulty, the unit price should perhaps be capped temporarily.

■ Subsidy reform will undoubtedly result in tariff increases. To gain acceptance from consumers, the increases should be accompanied by significant improvements in the reliability, quality and accessibility of electricity supply. Restoring the investment capability of SEBs – or their unbundled sub-divisions – should be a priority. An active communications programme to explain the rationale behind subsidy reform and market pricing must accompany the reform.

■ Cost-based electricity pricing needs to be implemented for all users. This requires an accurate data collection system and information on costs. If policy-makers find it appropriate to maintain partial subsidies for a particular category of consumers, the mechanism should be transparent and carefully monitored. The subsidy should also be allocated directly from the state budget to avoid burdening the utilities. It should expire within a set time frame. Access to electricity for low-income households should be carried out through direct support, or by mechanisms such as lifeline rates (see Chapter 2)[1].

■ Demand-side management and load management should be more actively pursued at the state levels for all sectors, particularly industry and agriculture, to reduce peak-supply shortages and increase the cost-efficiency of the system. For such measures to be successful, metering and pricing policies based on daily demand profiles should be implemented.

---

1. This policy recommendation applies to schemes such as Kutir Jyothi, facilitating access to electricity to low income groups.

DISTRIBUTION

Issues   The distribution grid must be expanded.

An accurate database on consumption, including such items as technical losses, must be made available to all market players. No such database exists in India.

In its policies for electricity reform, the central government identified the need to separate distribution from other activities carried out by SEBs. But, few states have implemented this policy.

The commercial and the technical elements of distribution could be easily separated. Billing customers and taking payment could be handled by a separate entity from the one responsible for the distribution of electric power to consumers.

Recommendations   ■ State governments should expedite reforms in the distribution sector, separating distribution from other activities carried out by SEBs. This process could be monitored at the central level. States should be accountable for the performance of their public electricity systems, particularly revenue collection by distributors.

■ Distributing entities should be placed on a commercial footing and improved revenue collection should be rewarded. Best practices should be disseminated to other states. In the state of Orissa, the privatisation of distribution fell short of expectations. Changing the tariff methodology was insufficient to allow for increasing revenue collection. It is hard to say whether this was due to a lack of political support or to inappropriate regulation. The central government should rapidly provide financial incentives for states to create innovative institutional and regulatory structures that improve revenue collection.

■ Once the revenue collection has improved, but only then, corporatised public distributors should be privatised.

■ Managing the distribution sector efficiently requires appropriate information about electricity consumption. Accordingly, a consolidated consumption database should be created and made transparent through improved metering. The management of distribution should be monitored using criteria based on the optimal use of resources. Indicators should include sales per kilometre of distribution line, average capacity of sub-stations, final marginal cost per kWh distributed, and equipment turnover rates.

TRANSMISSION

Issues   Average electricity consumption per square kilometre is very low in India compared with averages in OECD countries. This may justify the development of distributed generation rather than centralised power generation. The substantial auto-production capacity that already exists in the industry may be connected to the grid and emerge as distributed generation in the years to come. However, the bulk of power generation currently uses a centralised grid and is fuelled mostly by domestic hydro and coal concentrated in specific areas. This is why transmission is a key component of India's electricity supply industry. For the time being, interstate transmission is dictated by supply-demand imbalances between states. For the last 20 years, public expenditures in transmission have not been commensurate with generation expenditures. This increased T&D losses – already considerably burdened by power theft – and reduced pooling of Indian power-generation resources, and thus led to an unreliable transmission system.

The concept of regional planning and operation was adopted in the 1960s. Five regions were identified. The current development of domestic resources (hydro and coal) for the Indian electric power system is a first step in the gradual integration of the regional systems into an India-wide power system. Plans to build large power plants, particularly if they are fuelled by liquefied natural gas (LNG) which itself needs large infrastructure, will make an integrated national power system even more necessary.

SEBs tend to favour state-level solutions to reduce power shortages. They prefer to add generation capacity instead of developing interstate and inter-regional electricity trade. This is largely due to the soft budgetary constraints on SEBs and their insufficient use of cost criterion in investment decisions.

A central transmission company (CTU), responsible for developing the interstate transmission grid and power exchange, was established in 1989 as a public enterprise. It is called POWERGRID Corporation of India. The CERC is now formulating an interstate transmission tariff and a grid code.

For the time being, transmission pricing, operation and investment at the state level remain the responsibility of SEBs.

Recommendations

■ Given current losses in the transmission sector, investments in transmission are likely to be far more cost-effective than investments in generation. They should be given the highest priority by the central government. As a means to achieve a national electricity market, POWERGRID should be given a clear mandate and adequate capital to set up a national transmission grid. Private-sector participation should be encouraged to supplement public efforts through specific investment schemes such as build-operate-and-transfer (BOT).

■ The central government's plans are essential to guide investments in transmission since the large development needs of the system entail high investment risks. But, unlike the existing plans which do not refer explicitly to cost as an optimisation criterion, future plans should use economic criteria extensively.

■ An independent central system operator could be assigned the responsibility for operation, maintenance and development of the very-high-voltage transmission network (400 kV and above, both inter and intrastate). This operator should co-operate closely with another entity responsible for merit-order dispatching of all Indian electricity generation, including auto-production.

■ In the longer term, consumers who are connected directly to the transmission grid should be free to buy from any supplier.

■ Transmission pricing should anticipate the emergence of an interstate and intrastate competitive power market. Generally speaking, this pricing should be based on the following principles:

- the transmission system operator should provide access to the grid without discriminating among types of users;

- there should be no discrimination among customers when connecting new customers to the transmission network;

- use-of-system charges should not restrict, distort or prevent competition in the generation, supply or distribution of electricity.

GENERATION

Issues

In 1999-2000, of total Indian generation capacity amounting to 113 GW, 15 GW were auto-production[1]. Of the remaining 98 GW, coal-fired generation accounted for 61%, hydro 24%, gas 10%, nuclear 3% and oil 2%. Given India's vast coal resources, and its large untapped hydroelectric potential, these two resources are likely to provide the bulk of additional generation capacity in future.

Almost two thirds of the generation capacity in India is owned and operated by the states through electricity boards or electricity departments. Despite the opening of generation to IPPs in 1991, the private sector provides less than 10 GW of total generation capacity. The capacity of central generating companies has developed rapidly since their corporatisation and now represents around one-quarter of total capacity. These large power plants allocate their supply to more than one state through the interstate transmission grid. Bulk-power exchanges are still limited to supply from central generating units and surplus power exchanges from one state to another. Generation capacity is not centrally dispatched. Individual state power markets are too small for true competition among large IPPs since they cannot absorb large new generation additions rapidly.

Existing generation suffers from several recurrent problems. The efficiency and the availability of the coal power plants are low by international standards. A majority of the plants use low-heat-content and high-ash unwashed coal. This leads to a high number of airborne pollutants per unit of power produced. Moreover, past investments have skewed generation toward coal-fired power plants at the expense of peak-load capacity. In the context of fast-growing demand, large T&D losses and poor pooling of loads at the national level exacerbate the lack of generating capacity.

---

1. Often called captive production in India. In India, investments in auto-production are made primarily for standby purposes or as a substitute to electricity provided through the grid.

Making it possible for private investors to develop IPPs was one of the first elements of the liberalisation process initiated by the central government in 1991. The IPP policy has met with only mixed success in India:

■ about 250 projects were identified, mostly in memorandums of understanding with state authorities early in the liberalisation process. Most of the projects never reached the competitive-bidding stage and sites were not prioritised. Disputes over these projects have hampered the development of newer, more viable ones;

■ the time and effort required to develop new IPP projects in India proved too great for some foreign investors, and many reduced their Indian exposure. Among recurring difficulties are the lack of prior clearance of the projects by the authorities, problems in securing fuel supply agreements and the bankruptcy of SEBs;

■ the lack of prior prioritisation of the possible projects has led investors to perceive a high commercial risk in an overcrowded market-place.

Furthermore, the IPP policy has done little so far to increase the availability of electricity and reduce costs. In 2000, the Ministry of Power created a special group to review transparency in the bidding process. Lack of transparency has resulted in power purchase agreements (PPAs) that did not always meet the objective of increased power availability at lower prices. The only IPP projects that came online, apart from Dabhol (see Chapter 3), were small in scale, because of the limited size and the insolvency of the states' power markets. Many of these IPPs use naphtha, a costly oil product, as a fuel. In addition, the high commercial risk perceived by investors increases interest rates. Only about a dozen projects generating around 3,000 MW came online.

Public enterprises, such as the National Thermal Power Corporation, represent an increasing share of incremental capacity. Facing financial difficulties, the SEBs slowed their investments in incremental capacity in the 1990s and the public sector's withdrawal from generation has not been compensated by private investments. The gap between supply and demand has worsened.

Interstate sales of bulk-power are priced using a cost-plus mechanism. An availability based tariff (ABT) is now under discussion led by the CERC, and could be implemented in the near future (see Chapter 3). The tariff is intended to deal with current issues facing the power system, such as:

■ improving the availability of generation units;

■ penalising unscheduled interchanges, and;

■ establishing a level playing field for merit-order dispatching to be applied to state-owned capacity versus bulk-power supplied by other states or by central generating companies.

This tariff was devised as a first step toward a competitive bulk-power market. The efficiency of the device is questionable because the ABT is a relatively complex tariff mechanism. It will be difficult however to avoid such an intermediate step toward a fully-competitive power market. Electricity-market institutions are still in their infancy in India, and international experts agree that the transition to a fully-competitive market is a long-term process involving gradual changes.

A number of the difficulties in the generation link of the power chain will gradually be resolved as end-user payment for electricity improves. If it happens, this will boost the financial flow, improve the solvency of the purchasing utilities and reduce the commercial risk perceived by private investors.

Recommendations

- The time is now ripe for integration of the generation mix on an India-wide basis.

- Future plans should use economic criteria and least-cost utility planning to determine the optimal electric power system at the country level. Due consideration should be given to India's large-scale hydroelectric potential and sizeable coal resources, to the emergence of a market for imported natural gas and imported coal, to the need to keep the nuclear option open, and to the environmental effects of various technologies. The existence in neighbouring countries of a potential for surplus power from hydroelectricity or natural gas should also be considered. Development plans for coal and electric power should be co-ordinated to decide whether to transport coal or electricity. The gas industry should also be consulted, as it seeks to develop the gas network.

- The central government should facilitate investments by national generating companies and promote competition among them.

- Generation dispatching should be carried out at the national level. Dispatching is already done regionally through several mechanisms including Regional Load Dispatch Centres (RLDC), but it should be further implemented at the national level.

- The central government should use appropriate tariffs to accelerate the implementation of a framework for dispatch of surplus electricity to the grid from auto-producers.

- Generation projects for state markets should be limited in size. They should concentrate on renovating existing state capacity and on the developing of peaking capacity (combustion turbines and peaking hydroelectric power plants) and grid-connected renewables (such as wind and solar).

- All these measures highlight the need for further integration of the power generation sector on a national basis. There is a clear need for better co-ordination among the various ministries in charge of energy matters. National integration of the power generation sector and co-ordination among the various energy sectors will do better if technical, economic and environmental criteria outweigh political criterion in the decision process.

- The risk of market concentration at the state level calls for the development of a national market for generators. The creation of this power market at the national level requires an agreement on the bulk-power tariffs. Accordingly, the discussions on the ABT should be accelerated and its implementation expedited.

■ India could become more attractive to foreign investors if some risks were diminished:

- every IPP project should undergo a transparent bidding procedure. Projects should be better prepared before calls for tender are issued. Fuel supply agreements are crucial and should be secured in advance;

- contracts, once signed, must be upheld;

- commercial risk could be mitigated by prioritising projects at the central government level and by providing state guarantees through organisations such as the Power Trading Corporation (PTC);

- tariffs orders should be implemented over a sufficiently long period of time to offer a stable environment to investors.

■ India could benefit more from IPPs by rapidly making the bidding processes more transparent and fully competitive to avoid a later inflation of project costs.

■ Competition at the level of equipment supply should also increase. India should foster international technology co-operation to increase access to cheaper, cleaner and more efficient power technologies.

RURAL ELECTRIFICATION

Issues

A large part of the Indian population lives in rural areas. According to official statistics, most villages are electrified[1]. However, few households in these villages actually have access to the electricity grid[2].

The investment required to connect the remaining households to the main electricity grid is very large. Population density is low in rural areas, and a large part of the rural population has low income. Most Indian villages already have a main connection to the grid. The feeder – or medium-voltage line – that links the village to the network may actually be a very small part of the investment required to supply electricity to the households of a given village.

Electricity consumption for agriculture in rural zones is heavily subsidised. The average price paid by the agriculture and irrigation sector is reported to be equivalent to USD 0.5 cent/kWh. This was only 12.5% of the unit cost of supply. Low revenues from agriculture consumers limit the incentives for SEBs to develop consumption and to ensure good-quality power supply in electrified villages. Some consumers seeking more reliable service simply generate their own electricity.

---

1. 86% of villages in 1997, representing 75% of the total population.
2. 31% of the 112 million households living in rural zones benefit from electricity (1991 census). The definition of an electrified village does not account for the number of household connected but just the fact that an electricity line extends to that particular village.

Recommendations

■ Reforming subsidies and improving the payment rate for electricity already delivered are as important for rural electrification as for the generation, transmission and distribution sectors overall.

■ Much of rural electrification could take place outside the main grid, using decentralised supply or small-scale local grids and alternative energy sources such as biomass and wind. These solutions may require the creation of large credit facilities to stimulate investment.

■ The Government of India should encourage and experiment with innovative institutional models to supply electricity to rural consumers, such as co-operatives.

II. INTRODUCTION: THE ELECTRIC POWER MARKET IN INDIA

ELECTRIC POWER DEMAND IN INDIA

From the time of India's independence in 1947, the demand for electricity has grown rapidly. Final consumption of electricity has increased by an average of 7% per year since 1947. This sustained growth is the result of economic development and the increase in electrical appliances. It has been accompanied by a gradual shift from non-commercial sources of energy, such as biomass, in the household and commercial sector as well as the reduction in the use of coal for process heat in industry and kerosene for household lighting.

Of total final sales of 332 TWh in 1999-2000[1], industry accounted for just over one-third, agriculture for 30%[2] and the household sector for 18%. But for many years, electricity supply has fallen short of demand and the sustainability of this trend is very uncertain. Though the overall demand-supply gap decreased from an estimated 8.1% in 1997-98 to 5.9% in 1998-99, it rebounded to 6.2% in 1999-2000. Peak-power shortages fell from 18% in 1996-97 to 12% in 1999-2000.

In spite of sustained growth, electricity consumption per capita was only 416 kWh per annum (in 1998), far below the world average of 2,252 kWh.

The International Energy Agency's *World Energy Outlook 2000* projects electricity demand in India to increase by 5.4% per year from 1997 to 2020, faster than the assumed GDP growth rate of 4.9% (IEA, 2000a).

The duration and number of blackouts and brownouts are beyond acceptable limits, leading to shortfalls of up to 15% of demand. Consumption is largely constrained by the supply as Figure 2 shows. Figure 2 represents load charge in a large Indian city. As this figure shows, it seems that the seasonal variation in the load between Summer and Winter is limited, and changes during the day are not high either as compared to other countries. Inadequate power transmission and distribution result in shortages which in turn affect consumption patterns and induce commercial users and the most affluent domestic customers to rely on standby/in-house investments in auto-production capacity. Because of unsatisfied demand, increases in electricity prices would not

---

1. All statistics in India are based on the fiscal year that runs from April 1 to March 31. 1999-2000 sales are the sales recorded from 1 April 1999 to 31 March 2000.
2. A significant proportion of consumption reported as being in the agriculture sector is actually consumed by other sectors but not properly metered. Actual farm consumption could be only 10% of overall consumption. The remaining unrecorded 20% is considered in this study as non-technical losses.

Figure 1   Electricity Consumption in India

1971   **Total: 51 TWh**

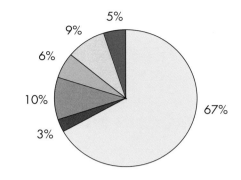

☐ Industry	▨ Commercial and Public Services
■ Transport	☐ Residential
▨ Agriculture	■ Other Sectors

1998   **Total: 376 TWh**

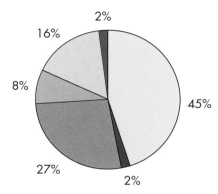

Source: IEA.

automatically lower consumption. A part of the population could afford a costlier electricity service, but available supply cannot satisfy their demand.

A large majority of the population is rural. Officially, close to 90% of the villages are electrified. However, only half the Indian population does in fact have electricity. Since a large portion of the population lives below the poverty line, they cannot afford electricity at current costs; this is particularly true in rural areas.

Table 1   Household Access to Electricity in India in 1997 (%)

Total access	45.7
Rural access	33.1
Urban access	81.5
Rural population as % of total population	74.0
Urban population as % of total population	26.0

Source: United States Energy Information Administration, World Bank.

Figure 2

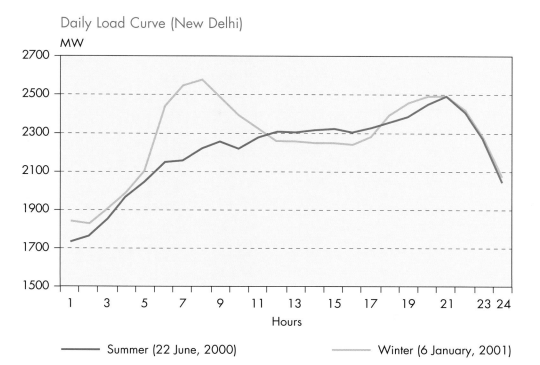

Daily Load Curve (New Delhi)

Source: Delhi Vidyut Board.

ELECTRIC POWER SUPPLY IN INDIA

India had 97,837 MW of generating capacity on 31 March 2000. In addition to this utility-owned capacity, a substantial amount of auto-production capacity exists mainly in the industrial sector, now amounting to around 15,000 MW, according to official data. Capacity additions of 4,242 MW were made during 1998-99. Growth in power generation has increased rapidly in recent years, from 301 TWh in 1992/3 to 451 TWh in 1998/9 – an average annual rate of growth of just over 6%.

India's electricity supply is mainly based on coal burnt in boilers feeding steam turbines, an adequate technology for baseload power generation and to a lesser extent for intermediate-load generation. Hydroelectricity capacity and the shares of hydro in generation have been decreasing over the years, reducing the availability of peak-load power. The Indian system is biased toward the production of baseload power, while the supply-demand gap is mainly in peak-load electricity. Gas is gaining an increasing role. Nuclear accounts for a marginal share of capacity and is not expected to be a major source of power in the immediate future.

India's electricity supply has long been dominated by the public sector. Public ownership, and public management of the main elements of the supply industry have been the rule since independence. At that time, most existing electric utilities were integrated into 19 SEBs (see Electric Supply Act of 1948) and eight electricity departments. These boards were part of state governments.

Figure 3   Electricity Generation by Fuel

1971   **Total: 61 TWh**

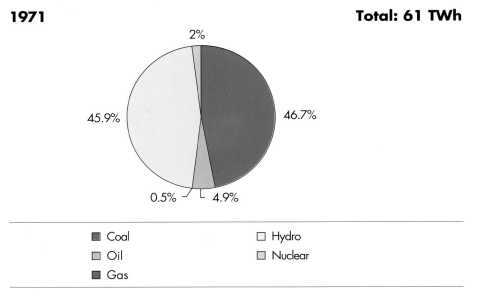

■ Coal	□ Hydro
■ Oil	□ Nuclear
■ Gas	

1998   **Total: 442 TWh**

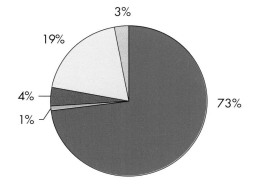

Source: IEA.

In 1998-99, the SEBs owned around 63% of generation capacity. The rest was owned by central sector utilities (CSU), such as the National Thermal Power Corporation (NTPC) or the Nuclear Power Corporation of India Ltd. (NPCIL; see Annex 2). CSUs were created after 1975, under the Indian Company Act, with administrative control in the hands of the central Ministry of Power. They were designed to pool state resources, such as hydroelectricity and coal, thus providing economies of scale, and to complement the SEBs' limited investment capability[1].

Fixed portions of the electricity generated by these national generation companies are allocated to states. Through its ownership of the Power Grid Corporation created in 1984, the central government has exclusive responsibility for high-voltage interstate transmission, which represents a small but growing share of total transmission.

---

1. This limited investment capability is primarily due to the fact that the SEBs, as part of the administration, are allocated a budget and cannot exert financial leverage by borrowing from commercial banks. In this regard, the corporatisation of the SEBs would greatly enhance their ability to invest.

Map 2 Main Power Plants

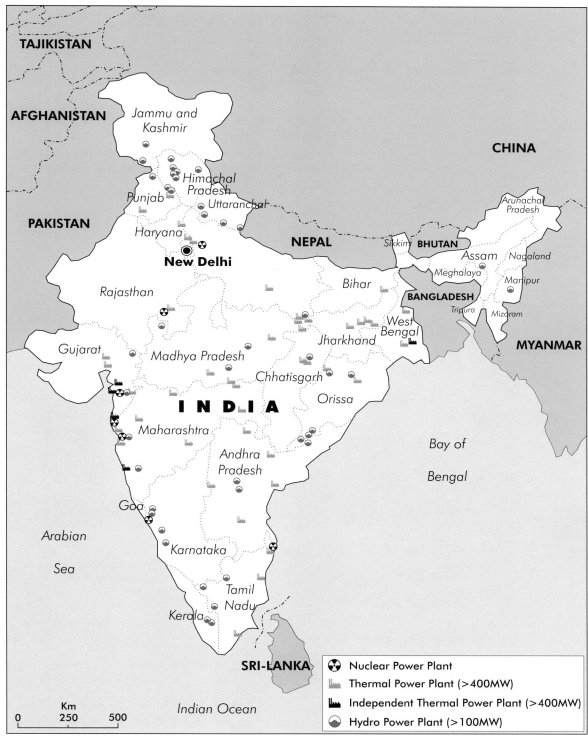

Source: TERI, 2000.

Transmission within states and most local distribution are in the hands of the SEBs. Until recently, all generation, transmission and distribution of power belonged to the public sector except for some licencees such as Bombay Suburban Electricity Supply (BSES), Tata Electric Corporation (TEC), or Calcutta Electricity Supply Corporation (CESC). Private utilities and independent producers represent only a marginal share of electricity supply. Since 1991, the sector has been open to private investors, initially for generation and later for transmission and distribution.

Capacity additions through public investments have fallen below the government's target, due to the large public deficit which has hampered the government's ability to invest. More generally, public investments in the power sector have not been commensurate to the needs in the past decade and have suffered from a bias toward the generation of electricity, rather than the transmission and distribution of power. The latter has been corrected in recent years.

Figure 4   Transmission and Distribution to Generation Ratio of Public Investments (Power-sector Plan Outlay), Billion Rupees

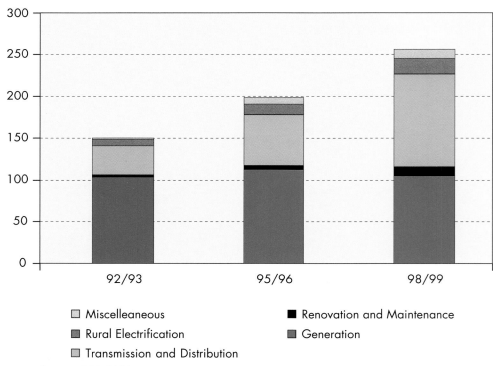

Source: GOI, 2001a.

Transmission and distribution losses due to the inadequacy of the system are very significant, varying between 20 and 45%. In terms of sales, this means that for every kWh of net generation, between 0.8 and 0.55 kWh is billed. In most OECD countries, the loss rate is less than 10%.

According to official estimates, roughly one-fourth of T&D losses is in transmission and three-fourths are in distribution. Average T&D losses seem to have increased over the past decade, mainly as the result of more realistic assessments by the states. The total figures of T&D losses may or may not include theft, non-metered and non-billed

Figure 5 1997-1998 Balance of the Indian Electric Power System, Millions of kWh

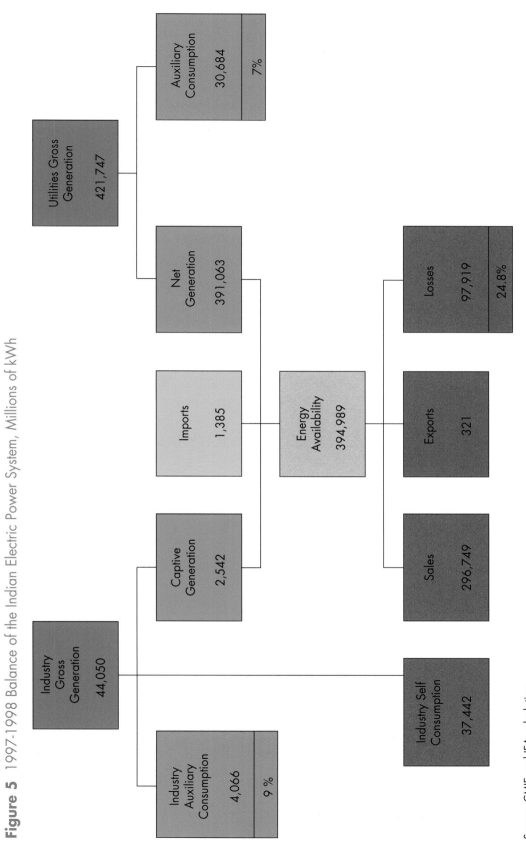

Source: CMIE and IEA calculation.

consumption. Due to the absence of adequate metering, these figures are inherently inaccurate and can lead to serious difficulties in determining tariffs[1].

The electricity-supply industry in India is characterised by a large amount of auto-production (often referred at as captive generation in India) developed over the years by energy-intensive industries such as steel and aluminium. Auto-production represented close to 15,000 MW in 1999-2000 and was increasing rapidly. Auto-production is a result of the high prices paid by industry consumers and the poor quality of electricity available from the network.

The efficiency of the overall electricity system is not high. Measured as the plant load factor in state-owned power plants, this indicator has not improved substantially over the past decade. In the transmission part of the system, several large failures of the grid occurred in recent years. A failure on 2 January 2001 blacked out all of North India for more than a day.

CO_2 emissions from the power sector represent half of total Indian emissions. They reached 399 million tonnes of CO_2 in 1999, out of the 904 million tonnes of CO_2 of total emissions. CO_2 emissions from public electricity and heat production grew at 8.2% per annum, or 93%, since 1990; against 53% for the total emissions. Steam power plants using coal with high ash content and low calorific value have long been identified as major contributors of airborne pollution (Wu & al., 1998).

LIBERALISATION POLICY: TOWARD AN ELECTRICITY MARKET

Energy policies are designed by various ministries: Ministry of Coal, Ministry of Petroleum and Natural Gas, Ministry of Power, Ministry of Non-Conventional Energy Sources, and Department of Atomic Energy. The Planning Commission provides an overall view (see Figure 6).

Under the Seventh Schedule of the Indian Constitution, the power sector is on the "concurrent list", meaning that state legislatures and the Parliament both have power to legislate in the sector. In the event of a conflict, Central Law prevails. A state can however legally enact conflicting legislation with the assent of the President of India.

The need to expand the implementation of market principles in the power sector and to stimulate the development of a power market with private players became obvious in the late 1980s, as the financial situation of the public players deteriorated and demand continued to go unmet.

Prices needed to be rationalised to increase energy efficiency, stimulate investments and ultimately reinforce energy security. Until recently, energy policy-makers controlled

---

1. For instance in Andhra Pradesh, T&D losses for financial year 1995 were reported to be 5,575 GWh (18.9% of net generation) and sales to agriculture were reported to be 11,757 GWh. The following year, T&D losses were 10,589 GWh (33.1%) and sales to agriculture were 8,210 GWh. In this case, it is clear that non-technical losses were mostly allocated to agriculture sales statistics. This had a drastic effect on the average tariff for agriculture, which was 2.8 paise/kWh in FY 1995 and surged to 13.4 paise/kWh the following year.

Figure 6 Organisation of the Power Sector in India

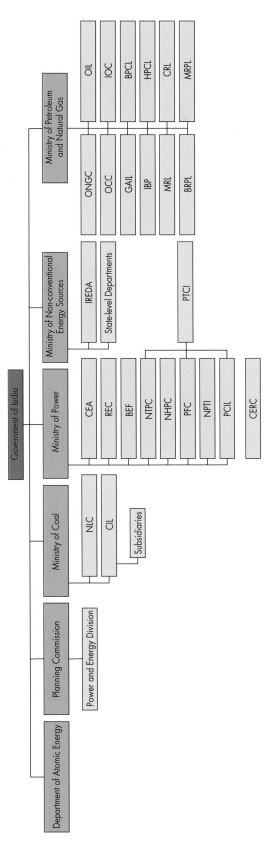

BPCL: Bharat Petroleum Corporation Ltd
BEF: Bureau of Energy Efficiency
BRPL: Bongaigaon Refinery and Petrochemicals Ltd
CEA: Central Electricity Authority
CERC: Central Electricity Regulatory Commission
CIL: Coal India Limited
CRL: Cochin Refineries Ltd

DAE: Department of Atomic Energy
GAIL: Gas Authority of India Ltd
HPCL: Hindustan Petroleum Corporation Ltd
IOC: Indian Oil Corporation Ltd
IREDA: India Renewable Energy Development Agency
IBP: Indo-Burma Petroleum Company Ltd
MRL: Madras Refineries Ltd

MRPL: Mangalore Refinery and Petrochemicals Ltd
MNES: Ministry of Non-conventional Energy Sources
NHPC: National Hydro-electric Power Corporation
NLC: Neyveli Lignite Corporation
NPTI: National Power Training Institute
NTPC: National Thermal Power Corporation
OCC: Oil Coordination Committee

OIL: Oil India Limited
ONGC: Oil and Natural Gas Corporation
PCIL: Powergrid Corporation of India ltd
PFC: Power Finance Corporation
PTCI: Power Trading Corporation of India ltd
REC: Rural Electrification Corporation

Source: TERI, 2000 and IEA.

all energy prices in India. Sometimes they actually defined who should do what in the market. Before the policy reforms, no independent authorities existed to regulate markets and facilitate competition. The main companies operating in the Indian electricity markets remained largely publicly-owned and had limited freedom to react to changes in domestic or international market conditions. Their role was pre-determined by public policy choices. The latter guided their production decisions, production levels and often the economic conditions of their operations, through control of their technology choices, the cost of capital and the cost of labour.

Recognising this situation, in the early 1990s, the government promoted the liberalisation of the electricity sector, envisaged along with the opening of the economy.

Measures to facilitate market development have been initiated by the Government of India (1991: opening the market to IPPs) and supported by multilateral organisations (1992: the World Bank "decided to lend only to states that agreed to totally unbundle their SEBs, to privatise distribution, and to facilitate private participation in generation", World Bank, 1999).

A significant commitment to regulatory and price reforms was made by the Common Minimum National Action Plan for Power, issued in December 1996 by the Chief Ministers of all Indian states led by the central Government of India. In this plan, the following guidelines were adopted to facilitate market pricing (a state responsibility):

"Each state/union territory shall set up an independent State Electricity Regulatory Commission (SERC).

(...)

To start with, such SERCs will undertake only tariff fixation.

Union government will set up a Central Electricity Regulatory Commission (CERC).

CERC will set up bulk tariffs for all central generating and transmission utilities."

And under Section IV. Rationalisation of Retail Tariffs:

"Determination of retail tariffs, including wheeling charges etc., will be decided by SERCs which will ensure a minimum overall 3% rate of return to each utility with immediate effect.

Cross-subsidisation between categories of consumers may be allowed by SERCs. No sector shall, however, pay less than 50% of the average cost of supply (cost of generation plus transmission and distribution). Tariffs for agricultural sector will be no less than fifty paise per kWh to be brought to 50% of the average cost in no more than three years.

Recommendations of SERCs are mandatory. If any deviations from tariffs recommended by it are made by a state/union territory government, it will have to provide for the financial implications of such deviations explicitly in the state budget."

This plan provided for some rationalisation of tariffs, notably that they should ensure a minimum overall 3% of rate of return, and that, after three years, no tariff should be under 50% of the average cost of supply. So far, no improvement in cost recovery from sales has occurred. In fact, the situation has worsened.

Map 3 High Voltage Transmission Grid, Above 400 kV

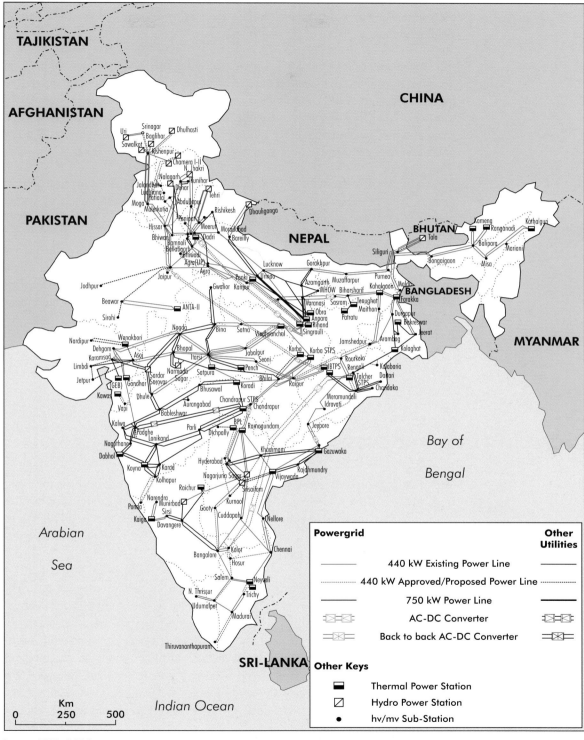

Source: TERI, 2000.

Implementing economic reforms contributed to higher GDP growth rates (an annual average close to 7% in the second half of the 1990s). Population growth of close to 2% a year added to income growth and boosted energy demand at rates slightly above GDP growth.

But implementation of much-needed electricity reforms has been slow, and crucial decisions have often been delayed or watered down. The two most important reasons for these delays are lack of co-ordination, or outright competition, among the numerous ministries in charge of energy matters in the central government and resistance to change on the part of sections of the Indian population which benefited from the energy policies of the past, reducing the scope for implementation in the states.

Politically sensitive decisions, such as the reduction of price subsidies for electricity to farmers, have been postponed or even cancelled, sending negative signals to market players and hampering market development.

Many states have been slow to implement reforms. In an attempt to boost the pace of implementation by the states, the central government tried to build a wider consensus on reforms at the Conference of Chief Ministers in March 2001. Among the decisions taken then was a commitment to work for the elimination of theft and the achievement of commercial viability in distribution[1].

A decade later, a full-fledged power market is still far from operational, but the reform policy has paved its way.

ELECTRICITY PRICES AND SUBSIDIES

More than any other factor, the way electricity prices are determined has inhibited India's power market development. Underpricing and political interference in price determination have worsened the financial situation of the main electricity producers, wholesale buyers and suppliers: the SEBs. This increases the risk for private players who wish to enter the electricity market.

The SEBs' end-use electricity tariffs vary widely according to customer category. The major categories are households, agriculture, commercial activities, industry and railways. There are large cross-subsidies between customer categories in India: tariffs for households and agriculture are generally well below actual supply costs, while tariffs to other customer categories are usually above the utilities' reported average cost of supply. In 1999-2000, the average price of electricity sold amounted to 208 paise/kWh – 26% below the average cost of supply (see Figure 7). According to official data[2], the total under-recovery of costs – the difference between total costs and total revenues – amounted to 272 billion rupees in 1999-2000, an increase of 190% since 1992-93. Most of this subsidy is reported to be for the agricultural sector. Cost recovery rates

---

1. GOI, 2001.
2. GOI, 2001a.

Figure 7   Average Tariffs, 1999-2000, paise/kWh

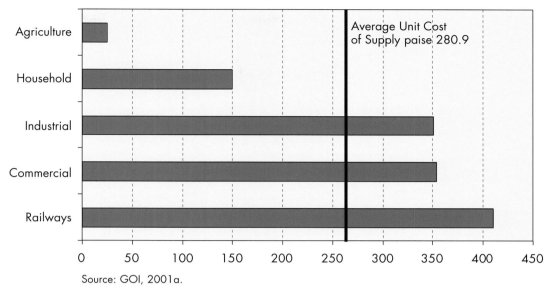

Source: GOI, 2001a.

vary markedly across the country. Rates are lowest in Jammu and Kashmir, Assam, and Bihar, while the highest rates (over 90% of cost recovery) are in Himachal Pradesh, Maharashtra and Tamil Nadu.

Figure 7 clearly demonstrates that subsidies are largest in the farm and household sectors, which are cross-subsidised by above-cost tariffs for commercial and industrial customers. A cost-reflective tariff structure would normally result in the lowest tariffs being charged to industrial customers, which have the highest consumption and load factor. The highest tariffs would be paid by household customers. Official data show that the nominal value of total subsidies to household customers quadrupled to 80.8 billion rupees from 1992-1993 to 1999-2000. Subsidies to agriculture more than tripled to 227 billion rupees over the same period.

The problem of underpricing worsened progressively through the early 1990s, to the degree that average revenues covered less than 76% of average costs by 1995-96. The recovery rate improved slightly up to 1997-98, but dropped sharply in the next two years to under 74%. This has mainly been due to a decline in average tariffs to agriculture, to under 25 paise/kWh. This has happened in spite of the introduction in Haryana, Himachal Pradesh, Orissa, Uttar Pradesh and Meghalaya of a minimum rate of 50 paise/kWh, as called for in the 1996 *Common Minimum National Action Plan for Power*. That plan also called for all end-use sectors ultimately to be charged no less than 50% of the average cost of supply, and within three years for agriculture. In no state has this goal been achieved.

Underpricing stimulates over-consumption by the beneficiary of the subsidies. In India, this is reflected in the excessive amount of electricity consumed by agriculture. Given the size of the population engaged in agriculture and its electoral importance, this largesse by policy-makers is difficult to eliminate. Vested interests hamper power-market development.

A review of historical data shows that consumption of electricity by agriculture multiplied by 19 from 1971 to 1998, whereas overall consumption multiplied only by seven. As of 1998, the sectoral structure of Indian electricity sales has differed dramatically from that of other Asian countries. In Asia, the domestic and commercial sectors generally account for almost half of electricity consumption and agriculture represents only 2%[1]. Industry accounts for 43% of overall consumption. The Indian figure is 45%, but more than one-third is auto-generated.

The amounts disbursed in subsidies are partly covered by cross-subsidies, which in turn burden less-favoured consumers, like industry. In 1999-2000, the prices paid by domestic customers, 149 paise/kWh (3.3 US cents/kWh[2]), and by customers registered as being from the agriculture sector, 25 paise/kWh (0.6 US cents/kWh), were far below the overall average price (208 paise/kWh). This occurred at the expense of commercial consumers (354 paise/kWh), industry (350 paise/kWh, 7.8 US cents/kWk) and railway hauling (411 paise/kWh).

The distortions go further. Since the average price per kilowatt-hour is calculated dividing the revenue collected by the quantity sold to a given category of consumer, official statistics probably underestimate the average price paid by agriculture, maybe by half. Indeed, by hiding non-metered or/and non-billed consumption from other sectors into the electricity sold to the farm sector[3], the average price is artificially deflated, and the actual amount of subsidies to the farm sector could be over-estimated. This is likely to blur the official appraisal of the amount of subsidies and their actual impact on consumption.

The large cross-subsidies from industrial, commercial, and railway hauling to the domestic and agriculture sectors tend to atrophy the paid consumption of the industrial and commercial sectors. Industry is subject to planned load-shedding[4], power cuts, voltage collapse and frequency variations, i.e. the high price paid by these customers is not compensated by good-quality supply. On the contrary, the poor quality of electricity service contributed to substantial industrial output losses.

The primary effect of underpricing is to distort the overall energy market in favour of electricity. Households, farmers and others who benefit from underpriced electricity consume as much cheap electric power from the grid as possible, and account for the bulk of demand. When the cheap central supply fails, private sources have to make up for the supply gap. Customers who need electricity supply invest in resources such as batteries and diesel generators. In so doing, they cannot benefit from the economies of scale arising from the grid and use systems that are not necessary very efficient in producing electricity. The double effect of underpricing, that has resulted in growing wastage of electricity over the past decades, and the development of auto-production largely explains why India has a higher electricity intensity of its GDP than the rest of Asia (excluding China; see Figure 8).

---

1. Excluding China, Korea and Japan.
2. This is at least ten times lower than the OECD average for the same category of customers.
3. For instance, in Uttar Pradesh, sales to agriculture for 1999-2000 were restated by the SEB as 5,122 GWh, versus 9,982 GWh resulting in an average price of 94 paise/kWh compared to the reported 48 paise/kWh (see World Bank, 2000).
4. The process of deliberately disconnecting pre-selected loads from the power system in response to a loss of power input to the system, in order to maintain the nominal value of the frequency.

Figure 8

Electricity Intensity of GDP in India and Asia, kWh of Final Electricity Consumed per Thousand 1990 USD

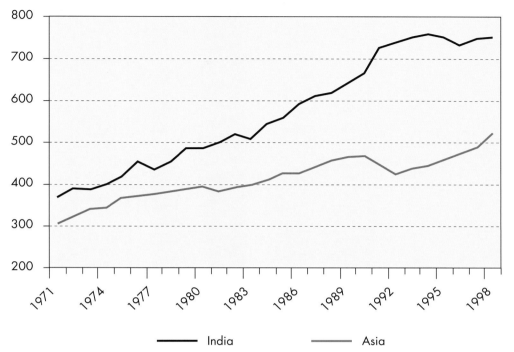

Note: Asia includes all non-OECD Asian countries, except China and India.
Source: IEA.

Central government and state budgets are burdened by subsidies, which account for a large share of current expenditures, at the expense of investments in the electricity sector or other sectors such as education and health.

The subsidy problem may be larger than or different from what is suggested by the data presented here. As we mentioned above, if most of these non-technical losses are allocated to sales to agriculture, the agriculture tariff issue could be less important than statistics make it appear. Conversely, the issue of non-payment could be more important than officially stated. Many customers, from agriculture but also households in urban areas, do not pay but continue to receive service. These customers effectively enjoy a 100% subsidy. This non-payment problem could far outweigh the official subsidies issue.

Since SEBs are managed like government, it is difficult to operate the power sector on the basis of economic criteria. Metering, billing and collecting revenues have been neglected. Decision-making remains highly centralised. Lower level employees have little decision-making authority.

An IEA study in 1999 attempted to quantify the size of electricity subsidies in India based on a price-gap analysis, the economic and (notional) financial cost of subsidies, and the potential impact subsidy removal would have on electricity consumption and related CO_2 emissions. That analysis has been updated with more recent price data (for 1999) and has been extended to cover agriculture. Table 2 summarises the results.

Figure 9   Average Electricity Supply Cost, Revenues and Cost-Recovery Rate

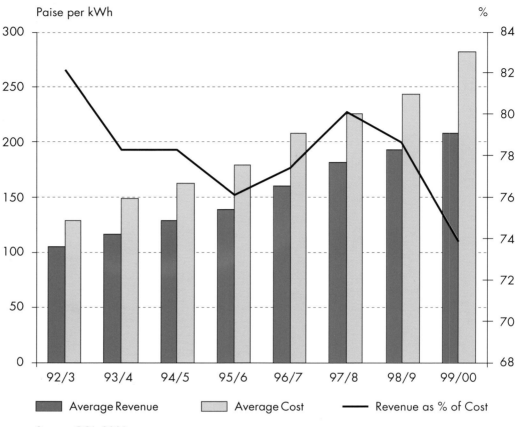

Source: GOI, 2001a.

Table 2   Electricity Subsidies: Summary of Results

	Average price (rupees/kWh)	Reference price (rupees/kWh)	Rate of subsidy (%)*	Potential primary energy saving from subsidy removal (%)**
Households	1.50	3.56	57.9	48
Industry	3.50	3.42	n.a.	0
Agriculture	0.25	3.56	93.0	86
Average	–	–	38.0	34

* Difference between actual price and reference price as percentage of reference price.
** TPES saved/TPES for the sectors covered by the IEA calculations.
Source: IEA calculation.

On the basis of 1999-2000 data, the rate of subsidy expressed as a proportion of the full cost-of-supply reference price amounted to 93% for agriculture and 58% for households. Electricity sales to industry are no longer subsidised. Using a –0.75 direct price elasticity of demand for the household, industry, and agriculture sectors, our analysis suggests that removing electricity subsidy would lead to significant reductions in electricity consumption, particularly in the agricultural sector. Total electricity use would be 40% lower in the absence of all subsidies. Assuming that the removal of subsidies on electricity sales reduces the demand for fuel inputs to power generation in equal proportion and that average thermal efficiency is constant at lower production levels, the use of coal and oil in thermal power plants would drop by 40%. This would lead to a 99-million-tonne reduction in power-sector CO_2 emissions, mostly due to lower coal use. The removal of specific coal subsidies would reduce emissions by an additional six million tonnes.

It is important to bear in mind the limitations of the price-gap approach, which identifies only static effects. It compares the actual situation with a hypothetical situation where there are no subsidies, keeping all other factors constant. In reality, this would never be the case. The dynamic effects of removing subsidies are likely to be significant. It could bring benefits in the form of greater price and cost transparency, gains in economic efficiency through increased competition and accountability and, consequently, accelerated technology deployment. These changes would offset, at least to some degree, the long-run static effects of subsidy removal on energy demand and related CO_2 emissions. This would be especially true for the electricity industry.

Subsidy reform, to the extent that it increases the SEBs' financial viability, would boost their capacity to invest and, therefore, increase sales to customers who currently lack access to electricity. In the long run, a reduction in subsidies could lead to an increase in electricity consumption by end-users not currently served or whose supply is severely curtailed, by blackouts, brownouts or time-limited service. Indeed, this is the implicit goal of electricity sector reforms, including subsidy reduction.

Figure 10   Electricity Prices and Consumption in India (1978-1996), Prices in Rs 1990 per kWh and Consumption in GWh

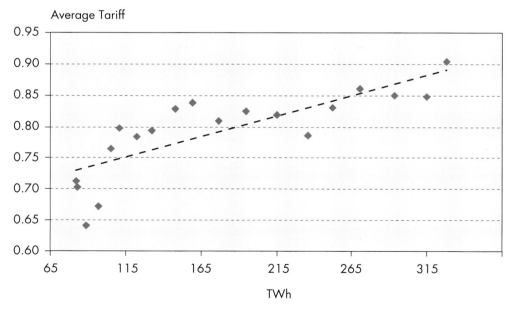

Source: IEA.

Whether this dynamic effect would be large enough to offset completely the static effect of higher prices is unclear. The actual size of the static effect is also unclear. As Figure 10 indicates, past experience shows a positive relationship between electricity prices and electricity consumed. The speed with which subsidy removal would lead to increased investment is also uncertain.

Any attempt to quantify the impact of electricity subsidy reform on investment in power generation would also need to take into account the economics of auto-production. One effect of the current structure of cross-subsidies is that investment has partially shifted from the SEBs to industry itself. The 1996 *Common Minimum National Action Plan for Power* sought to promote auto-generation by calling on SEBs to provide access to their grid to transmit power that is surplus to a company's own needs to other end-users.

III. POLICIES FOR A POWER MARKET AND RESPONSES FROM MARKET PLAYERS

Since the early 1990s, the Indian federal government has tried to introduce market mechanisms in the electricity-supply industry. These reforms were aimed primarily at meeting the need for additional investments in the power system which, given the large public deficit, would have to come largely from the private sector, domestic or foreign.

The objective was to pave the way for a full-fledged power market to emerge in the medium term. The Electricity Regulation Act of 1998 established the Central Electricity Regulatory Commission (CERC) with a mandate to set tariffs for interstate exchange and multistate generation. The Act also allowed for the creation of state electricity regulatory commissions (SERC), with powers to regulate retail prices.

In December 2000, after several years of debate, the CERC passed a Tariff Order to regulate prices charged to SEBs by power plants owned by the central government (Central Sector Utilities – CSUs). This tariff is intended as a step towards a competitive bulk-power market.

These measures elicited a positive reaction from private investors, who have initiated a number of generation projects or participated in public transmission projects. But the power sector is still hamstrung by the deteriorating financial situation of the SEBs and their mismanagement. In March 2001, the prime minister called state chief ministers and state power ministers to a meeting, also attended by the union power minister and finance minister, to discuss power-sector reforms. This meeting resulted in a decision to form a high-level group to work out a plan for the one-time settlement of the SEBs' debts to CSUs.

MAIN POLICY CHANGES

The government's main objective is to make power available and affordable to all. The main texts currently governing the electricity industry and the power market in India still are the Indian Electricity Act (1910) and the Electricity (Supply) Act (1948). We review below these legal documents and their amendments, as well the Electricity Regulatory Commissions Act (1998), other policy statements, additional resolutions and new regulations.

Indian Electricity Act, 1910

The Indian Electricity Act regulates the granting of licences to market operators: producers, transmitters and distributors of electricity. It defines who controls the distribution and consumption of energy. It regulates licencees' accounts, the installation of electricity-supply lines and other works. Finally, it determines who controls the supply, transmission and use of energy by non-licencees. The Act calls for fair treatment of customers and to do so, the Act requires that tariffs are defined through policy.

Electricity (Supply) Act, 1948

The Electricity (Supply) Act established the SEBs. SEBs do not have to comply with the financial and accounting rules of the Indian Company Act. They are vertically-integrated utilities with a commitment to enlarge the customer base from urban areas to rural areas. The Act defines the power and duties of the SEBs and generating companies. It defines the approval process for the establishment, acquisition and replacement of power stations. The Sixth Schedule of the 1948 Act established financial principles for determining licencees' tariffs, stipulating that profit should not exceed a 16% internal rate of return. The Sixth Schedule uses a cost-plus methodology in determining those returns.

Policy on Private Participation in the Power Sector, 1991

Under this policy, private-sector entrepreneurs may set up companies, either as licencees or as generating companies. Up to 100% foreign equity participation is allowed. The introduction to the resolution reads:

"With the objective of bringing in additionality of resources, for the capacity addition programme in the electricity sector, Government have formulated a policy to encourage greater participation by privately owned enterprises in the electricity generation, supply and distribution field. The policy, in this regard has widened the scope of private investment in the electricity sector, and has introduced modifications in the financial, administrative and legal environment, for the private enterprises in the electricity sector towards making investments in the sector by private units attractive. Based on this policy, a scheme has been framed to encourage private enterprises' participation in power generation, supply and distribution, ..."

Electricity Laws (Amendment) Act, 1991

This amendment of the Indian Electricity Act reinforced the integration of the grid in India by giving more authority to the regional load dispatch centres (RLDC).

Common Minimum National Action Plan for Power (Chief Ministers, December 1996)

In December 1996, the chief ministers recognised the need to reform the electricity-supply industry. The following extracts from the Plan indicate the intended direction of the reforms:

- *"the gap between demand and supply of power is widening;*

- *the financial position of State Electricity Boards is fast deteriorating and the future development of the power sector cannot be sustained without viable State Electricity Boards and improvement of operational performance of State Electricity Boards;*

- *reforms and restructuring of State Electricity Boards are urgent and must be carried out in a definite time frame;*

- *the creation of Regulatory Commissions is a step in this direction;*

- *the requirements of the future expansion and improvement of the power sector cannot be fully achieved through public resources alone and it is essential to encourage private-sector participation in generation, transmission and distribution;*

- *the changing scenario in the power sector calls for further delegation of powers and simplification of procedures;*

- *a national consensus has evolved for improving the performance of the power sector in a time bound manner and the following was adopted.*

I. National Energy Policy:

The Government will soon finalise a National Energy Policy.

II. State Electricity Regulatory Commission:

Each State/Union Territory shall set up an independent State Electricity Regulatory Commission (SERC).

To set up SERCs, central Government will amend Indian Electricity Act, 1910 and Electricity (Supply) Act, 1948. To start with, such SERCs will undertake only tariff fixation.

Licensing, planning and other related functions could also be delegated to SERCs as and when each state Government notifies it. Appeals against orders of SERCs will be to respective High Courts unless any state Government specifically prefers such appeals being made to the Central Electricity Regulatory Commission.

III. Central Electricity Regulatory Commission:

Union Government will set up a Central Electricity Regulatory Commission (CERC). CERC will set the bulk tariffs for all Central generating and transmission utilities.

Licensing, planning and other related functions could also be delegated to CERC as and when the central Government notifies it. All issues concerning interstate flow and exchange of power shall also be decided by the CERC. To enable setting up of CERC, central Government will amend Indian Electricity Act, 1910 and Electricity (Supply) Act, 1948.

IV. Rationalisation of Retail Tariffs:

Determination of retail tariffs, including wheeling charges etc., will be decided by SERCs which will ensure a minimum overall 3% rate of return to each utility with immediate effect. Cross-subsidisation between categories of consumers may be allowed by SERCs. No sector shall, however, pay less than 50% of the average cost of supply (cost of generation plus transmission and distribution). Tariffs for agricultural sector will not be less than fifty paise per kWh to be brought to 50% of the average cost in not more than three years. Recommendations of SERCs are mandatory. If any deviations from tariffs recommended by it are made by a state/UT Government, it will have to provide for the financial implications of such deviations explicitly in the state budget. Fuel Adjustment Charges (FAC) would be automatically incorporated in the tariff. There shall be a package of incentives and disincentives to encourage and facilitate the implementation of tariff rationalisation by the states.

V. Private Sector Participation in Distribution:

State Governments agree to a gradual programme of private-sector participation in distribution of electricity. The process of private participation shall be initially in one or two viable geographical areas covering both urban and rural areas in a state and the state may extend this to other parts of the state gradually.

VII. Autonomy to the State Electricity Boards:

States will allow maximum possible autonomy to the State Electricity Boards. The State Electricity Boards will be restructured and corporatised and run on commercial basis.

VIII. Improvements in the Management Practices of State Electricity Boards:

State Electricity Boards will professionalise their technical inventory manpower and project management practices.

X. Co-generation/Captive Power Plants:

State Governments will encourage co-generation/captive power plants. To facilitate evacuation of power from these plants to the grids, states shall formulate clear and transparent policies for purchase of power and wheeling charges which provide fair returns to the Co-generation/Captive power plant owners. Captive power plants could also sell power to a group of industries as well as other categories of consumers in the said industrial zone or area. Wheeling of power from captive power plants to consumers located at a distance or through displacement basis shall be encouraged and the states will issue clear and transparent long term policies in this regard."

Chief Ministers' Power Reform Initiative, December 1998

The Chief Ministers reiterated the need for co-ordination between SEBs, as well as the importance of distribution reforms.

Electricity Regulatory Commissions Act, 1998

This Act establishes a central independent regulatory commission and allows states to establish their own commissions. The CERC has a mandate to introduce competition and efficiency in the electricity-supply industry both centrally and in interstate operations. Tariffs, conditions of supply and service and, in many cases, licensing of investments and operations are within its purview.

The SERCs are a mirror of the CERC at the state level. Their primary role is to rationalise retail tariffs, but their mandate also covers the determination of wholesale[1],

---

1. I.e. from producers to dispatchers.

bulk[1] and grid[2] tariffs. They regulate power purchases made by transmission and distribution utilities, as well as trade among state generation, transmission and distribution utilities.

Electricity Law (Amendment) Act, 1998

Section 27(A) and (B) of the Indian Electricity Act, 1910 was modified to set up central and state transmission utilities. Transmission activity was given an independent status. POWERGRID was designated the central transmission utility (CTU) and the SEBs or their successors became state transmission utilities (STUs). The Act stipulates that the CTU and STUs are public companies. The CTU was given the responsibility for interstate transmission of electricity and for all planning and co-ordination of electricity dispatch.

Electricity Bill (8th Draft, September 2000)

By the end of 2001, this bill had not been passed and was still under discussion. It had initially been prepared by the National Council of Applied Economic Research (NCAER) for the Ministry of Power. The main aims of the bill are to consolidate all relevant legal texts in one piece of legislation, and to accelerate the reforms and restructuring necessary to ensure a healthy power industry using market principles. It rests on the idea that, despite several amendments to the 1910 and 1948 Acts, the basic concepts and structures embedded in these Acts remain unchanged.

The Bill recognises the present industry structure, based on state-owned monopolies, and creates a framework for the state to pursue a cautious yet definite process of restructuring and liberalising the electricity industry. The target set for the industry is to support an annual economic growth rate of 7-8%.

In particular, the bill provides for the development of regional and national grids to ensure electricity flows between regions. It also provides for the establishment of a bulk electricity spot market. In addition, the bill:

- reinforces the technical role of the Central Electricity Authority (CEA);

- commits the government to publish a national electricity plan;

- delicences generation and permit freely auto-production. Hydroelectric projects would, however, need approval from the state government and clearance from the CEA;

- provides for open access in transmission, with provisions for controlled surcharges to deal with existing cross subsidies with the surcharge being progressively phased out;

- allows distribution licencees freely to undertake generation and generating companies to take up distribution licencees;

---

1. I.e. from dispatchers to distributors.
2. I.e. transmission.

- encourages the government to adopt measures to promote rural electrification. Provides unrestricted licences for generation and distribution in rural areas;

- envisages setting up independent regional and state transmission centres for the non-discriminatory dispatching vital for IPPs;

- makes the Grid Code the tool to regulate load dispatching;

- retains all the basic features of the 1998 Electricity Regulatory Commission Act and entrusts the responsibility for introducing competition in the electricity sector to the regulatory commissions. This includes the establishment of pricing principles and the methodologies for the encouragement of competition;

- makes it mandatory for states to set up SERCs;

- requires the subsidy to be paid out of the state budget;

- acknowledges electricity trading as a distinct activity and mandates the regulatory commissions to regulate trading.

Chief Ministers Conference, March 2001

Faced with a growing financial crisis in the SEBs, and slow implementation of reforms at the state level, the central government called this meeting. The central government decided to support the states by formulating a one-time settlement of SEB dues to CSUs. The chief ministers re-iterated their commitment to work toward the elimination of power theft. They agreed to complete rural electrification by 2007 (the end of the 10th Plan) and to achieve full coverage of all households by the end of 2012 (the end of the 11th Plan).

The Energy Conservation Act, 2001

Passed in Autumn 2001, this Act visualises the setting up of a Bureau of Energy Efficiency to boost energy conservation activities such as implementation of standards and labelling for electric appliances, audits or awareness campaigns.

REGULATORY CHANGES AT THE FEDERAL LEVEL

The Central Electricity Regulatory Commission

The CERC was established in 1998 under the Electricity Regulatory Commissions Act, 1998 with a mandate to promote competition, efficiency and economy in bulk-power markets. It was also to improve the quality of supply, promote investments and advise the government on the removal of institutional barriers (see Box 1).

The CERC is expected to carry out the following functions:

- regulate the tariffs of generating companies owned or controlled by the central government;

- regulate the tariffs of generating companies other than those owned or controlled by the central government if such companies plan to generate and sell electricity in more than one state;

- regulate the interstate transmission of electricity including the tariffs of the transmission utilities;

- promote competition, efficiency and economy in the electricity industry;

- aid and advise the central government in the formulation of a tariff policy which will be fair to consumers and which will facilitate the mobilisation of adequate resources for the power sector;

- co-operate with the environmental regulatory agencies to develop appropriate policies and procedures for environmental regulation of the power sector;

- develop guidelines in matters relating to electricity tariffs;

- arbitrate or adjudicate disputes involving generating companies or transmission utilities regarding tariffs;

- aid and advise the central government on any other matter referred to the Central Commission by that government;

- issue licences for the construction, maintenance and operation of the interstate transmission system.

The CERC has established its authority over the power sector through a number of orders (listed in Table 3). These orders centre around the design and implementation of pricing rules for CSUs and sales from very large power projects, as well as a code to regulate interstate use of the grid.

Box 1   Mission Statement of the CERC

"The Commission intends to promote competition, efficiency and economy in bulk-power markets, improve the quality of supply, promote investments and advise the government on the removal of institutional barriers to bridge the demand/supply gap and thus foster the interests of consumers. In pursuit of these objectives the Commission will:

- improve the operations and management of the regional transmission systems through the formulation of Indian Electricity Grid Code and restructuring of the institutional arrangements thereof;

- formulate a tariff-setting mechanism, which provides speedy and time bound disposal of tariff petitions, promotes competition, efficiency and economy in pricing of bulk-power and transmission services and least cost investments;

- improve access to information for all stakeholders;

- institute mechanisms to ensure that interstate transmission investment decisions are taken transparently, in a participative mode and are justifiable on the basis of least cost;

- facilitate the technological and institutional changes required for the development of competitive markets in bulk-power and transmission services;

- advise on the removal of barriers to entry and exit for capital management, within the limits of environmental safety and security concerns and the existing legislative requirements, as the first step to the creation of competitive markets;

- associate with environmental regulatory agencies for the application of economic principles to the formulation of environmental regulations."

Table 3   CERC's Orders (as of June 2001)

General	
Purchase of power from MSEB and sale to Jindal	28 Oct. 1999
Grid code order	30 Oct. 1999
PGCIL IEGC order on review petition	21 Dec. 1999
Availability based tariff	4 Jan. 2000
Stay order – ABT order	7 Mar. 2000
Progress for implementation of IEGC	13 Jun. 2000
Regulation of power supply to central utilities	21 Jun. 2000
Review of IEGC dated 30.10.99 & 21.12.99	22 Jun. 2000
Tariff norms for central generating stations	21 Dec. 2000
PGCIL – Payment of fees and charges to RLDC for the year 1998-99	3 Jan. 2001
Review of progress for implementation of ABT	19 Mar. 2001

Generation Tariff	
Hydro	
NHPC – Loktak hydroelectric project	29 Feb. 2000
NHPC – Reducing the normative availability	29 Feb. 2000
NHPC – Kopili hydroelectric project	4 May 2000
NHPC – Kopili hydroelectric project for the year 2000-01	4 May 2000
NHPC – Ranjit hydroelectric project	12 May 2000
NHPC – Operational norms for hydro power stations	8 Dec. 2000
NHPC – ABT Review	8 Dec. 2000
Coal	
PTC – Pipavav mega power project	9 Mar. 2000
PTC – Purchase of power from Hirma	31 May 2000
PTC – Purchase of power from Hirma	13 Jun. 2000
PTC – Purchase of power from Hirma	21 Jun. 2000
NTPC – Incentives for 1998-99	23 Jun. 2000
NTPC – Secondary energy charges	17 Jul. 2000
PTC – Purchase of power from Hirma	3 Aug. 2000
PTC – Purchase of power from Hirma	26 Sept. 2000
NTPC – ABT Review	15 Dec. 2000
NLC – ABT Review	21 Dec. 2000
Gas	
NTPC – Kayamkulam CCGT	24 Jul. 2000
NTPC – Faridabad gas – 3/99	23 Aug. 2000
Transmission Tariff	
PGCIL – Kayamkulam – Pollam	17 Mar. 2000
PGCIL – Unchhar-Kanpur line – II at Kanpur	24 Apr. 2000
PGCIL – Korba- Raipur line out of Kanpur	24 Apr. 2000
PGCIL – Kaiga-Sirsi line	26 Apr. 2000
PGCIL – Incentives based on availability	19 Jun. 2000
PGCIL – Kaiga transmission system	4 Aug. 2000
PGCIL – Incentives based on availability	26 Sept. 2000
MPEB – Fixation of wheeling charges	23 Oct. 2000
PGCIL – Norms for interstate transmission	8 Dec. 2000

Source: www.cercind.org

The Availability-Based Tariff (ABT) and Wholesale Power Trade

Principles and Definitions

Two of the main tasks transferred to the CERC were to increase and optimise the economic efficiency of central generating companies and to establish an adequate price for electricity sold by central generating companies to their customers, the SEBs. Until the 1990s, tariffs for sales from generating companies to SEBs were decided by mutual consent between the supplier and the buyer. In 1992, to improve the efficiency of price determination, a committee was formed under the chairmanship of Shri K.P. Rao, which produced a new tariff schedule. The 4 January 2000 ABT Order, the 7 March 2000 Stay-ABT Order and 21 December 2000 Tariff Norms' Order of the CERC are the latest attempts to share the financial burden between central and state governments. They constitute a framework for notifications of terms and conditions for tariffs regulated under Sections 13(a), (b) and (c) of the ERC Act 1998.

The CERC views the Availability Based Tariff (ABT) as a transition toward fully competitive pricing[1]. Ultimately, the CERC thinks, supply and demand will determine the appropriate tariff. But this is far from the case at present due to the lack of complete transmission facilities and constraints in power trading. Therefore, CERC has endeavoured to develop a tariff mechanism, keeping in mind the interests of generating companies and customers while contributing to the development of a viable power sector in India (CERC, *Tariff Norms for Central Generating Stations*, 21 Dec. 2000, p. 157-158).

The 1992 tariff notification for sales by generating companies to the SEBs confirmed the use of a cost-plus tariff: a plant-based tariff in which normative costs are passed through to buyers. Among the norms used was one determining a minimum load factor for recovering full fixed charges (68.5% for steam coal units)[2]. The 1992 scheme introduced a two-part tariff, one part for the recovery of annual fixed charges, the other for the recovery of variable costs. It also introduced the concept of "deemed generation", under which a power plant fully recovers its fixed charges if it is available[3], even if it is backed-down.

The ABT is designed for the interchange of electricity between states or for supplies from central generating stations to state entities. It tries to move away from the cost-plus tariff approach. The ABT is designed to further some of the goals of the 1992 tariff by:

- promoting grid discipline so that the five electricity regions of India can operate in an integrated way, especially limiting and avoiding frequency fluctuations;

- discouraging surplus generation when frequency is high;

- creating a fixed charge varying with the generation capacity allocated to the customers. This is to level the playing field between central and state-owned power generation units by including all the fixed costs incurred by the central generation units in the fixed part of the tariff.

The ABT is a tariff for transactions between the operator of the power plant (or station) and the beneficiary. It consists of three parts:

---

1. In a competitive power market, plants sell electricity at a price defined by the supply-demand balance, corresponding to the short-term marginal cost. In such a system, plants are not guaranteed to recover their fixed charges. The Indian regulation aims to reduce the price volatility that could occur in a fully competitive power market, while providing additional incentives for more efficiency.

2. A load factor of 68.5% corresponds to yearly operation at full capacity of 6,000 hours per year. The 1992 tariff notification included some normative net heat rates (34% for steam coal units, 43% for combined cycle units and 30% for combustion turbines).

3. In the regulated Indian market, the definition of reference availability allows plants to recover fully their fixed charges. Availability means the quantity and time for which a given plant or unit commits itself to be available for power production. The availability of a power plant is defined as the percentage of the year when it is available for power generation at full capacity. Power-plant operators have to commit themselves to a generation schedule one day in advance. The so-called availability based tariff provides a formula in which the revenues of the plant operators vary with the availability of the plant, given a reference availability. The more available a plant is, the higher its revenue and the closer the actual availability is to the committed availability, the higher the revenue.

■ a fixed charge varying with the share of generation capacity allocated to the beneficiary and the availability level achieved by the generator. The tariff reflects the fixed costs incurred by the owner and operator of the plant. For hydroelectric power plants, 90% of the lowest variable generation cost of the thermal power plants in the region is deducted and added to the variable cost;

■ an energy charge based on daily scheduled supply;

■ a charge for unscheduled interchanges (UI charges).

After a detailed consultation process, financial and operational Tariff Norms were decreed in the Order of 21 December 2000 (CERC). The Order stipulates that its terms and conditions shall apply to all utilities covered under Section 13(a) (b) and (c) of the ERC Act. Its conditions shall be in force for a period of three years from 1 April 2001[1] and are reviewable at the discretion of the ERC. The unit for which the Tariff Norms are defined is the power plant for generation and the regional level aggregated line for transmission.

The Tariff Norms define the tariff base in terms of historical costs, that is observed past costs,[2] rather than the long-run marginal costs, which are the current and future costs of developing the system. The methodology uses a performance-based rate of return, rather than Retail Price Index minus-X, based on historical costs. The rate of return remains the same as before the Order, 16% on the debt of the power plant project.

The Order also deals with depreciation, on a straight-line basis, spreading the depreciable value over the useful life of the asset. There is however scope for reviewing the useful life of the asset. The Order defines the base for operation and maintenance (O&M) costs as the average normal actual O&M expenditure of the power plant or region. The escalation formula was made more realistic to reflect movement in the wholesale and retail price indices. The past five-year trend in indices is used to project O&M cost for the entire tariff period, thereby smoothing out the tariff impact. For new projects however, capital costs will be used as a basis. Norms are also defined for passing fuel costs through to customers, foreign-exchange rate variations, corporate taxation and incentives for exemplary performance. The Order also includes provisions for an additional development surcharge (5% for NTPC, NHPC, and NLC; 10% for POWERGRID).

Critique of ABT

The ABT represents some progress in three areas:

■ it levels the playing field between central public sector undertakings' power stations and the state's stations. The ABT introduces a fixed charge for a given power allocation by central generating stations to the states' utilities. The states determine their power purchases on the basis of the variable costs of both central and state generating power

---

1. Southern Region: 1 April 2001; Eastern Region: 1 May 2001; Northern Region: 1 June 2001; Western Region: 1 August 2001.
2. With the risk of under-investment reducing these costs.

stations. Previously, the states' variable costs were compared with the central stations' total costs;

■ it offers an incentive to central generators to abide by a dispatch schedule, facilitating the operation of the power system;

■ it gives the performance incentives to the central power plants.

The Indian electricity-supply industry desperately needs to attract private investment, domestic or foreign. The ABT provides a good and predictable framework for those operators of large, independent power projects who are willing to sell power to more than one state. Even though the central generating companies and most of the customers they supply will still be publicly-owned, they will be expected to operate more and more according to market principles. In this respect, the ABT may prove to be a valuable step towards a competitive bulk-power market.

It is too early to draw lessons from the ABT since it is yet to become fully operational nation-wide. However, a bulk-power tariff may be evaluated using the following criteria:

■ does it favour supply-demand adjustments in the short run as well as the long run?

■ does it provide incentives to reduce the overall cost of the commodity?

■ does it entail limited transaction costs?

■ is it predictable?

The ABT meets the last two criteria. Transaction costs will be reduced, since a number of sensitive issues were tackled during the detailed consultation process initiated by the CERC. Once the tariff has been finally defined, the CERC wants to limit further negotiations. A three-year period for implementation of the initial norms guarantees the predictability of the first tariff. The CERC favours an eventual five-year validity period for the tariff, which gives the market's private players plenty of time to adjust.

To minimise the overall cost of the commodity, the ABT should ideally result in the merit-order dispatching of each generating unit of the power stations, that is each turbine, allowing for the generating units' load-following and start-up costs. It should also provide power plants with an incentive to offer ancillary services to adjust their output to the grid's requirement (such as load/frequency adjustments, or the establishment of spinning reserves). It should give an incentive to reduce the variable operating and maintenance costs of each generating unit. The ABT partly meets these criteria. The tariff is defined for each power station. It does not price separately the variable costs of each unit. Since the National Thermal Power Corporation owns most of the power stations which will be under the ABT, it would be more efficient for the system if NTPC could decide at any moment which unit is to be dispatched.

Another issue is the limited scope of the tariff, which applies only for generating stations supplying more than one state or for central generating stations. IPPs which signed long-term PPAs before the ABT came into force may have an advantage over central generating stations since their price through the PPAs may be better than the one they would get under the ABT.

The ABT does not ensure that supply matches demand at any given point in time. A system operator would be required to handle overall co-ordination. The ABT does provide incentives to increase supply efficiency at the plant level. But the final cost of electricity sold to the SEBs will not reflect the real scarcity of electricity at country-level at any point in time, since most of the price components are fixed in advance. Fuel charges are largely fixed in FSAs and fuel allocations are defined in advance. There is no real way to compare costs between central stations and those belonging to states. Merit-order dispatching is not methodically practised at the regional level. The states operate their own units and may draw upon their allocated share from central generation units. Though the ABT rationalises the cost comparison between central stations and those of states, its scope is limited and an accurate cost comparison is still difficult to perform.

The ABT does not provide for overall least-cost operation of regional and national India generation capacity. Indian hydroelectric plants deserve specific treatment different from that scheduled in the ABT in order to maximise their usefulness in the system. The use of 90% of the least-variable cost thermal unit as a sole reference for all hydroelectric sources does not ensure that they will be optimally used. In theory, the price of hydroelectric power should vary according to the value of the water in reservoirs, which is based on the time period and the volume of stored water in the reservoir.

Since the measure of availability is one of the components of the tariff, and since the availability baseline for full recovery of fixed cost has been increased[1], the ABT may reduce the overall revenue of central generating companies compared to the previous situation. The difference between central generating companies and state companies is justifiable only if it does not last long. It is acceptable provided the ability of investment of the central generating companies is closely monitored to avoid major reductions in investment. It is also acceptable provided the tariffs used by the states also introduce rapidly and within a given period of time similar incentives for the performance of the state electricity generating plants. This difference has not gone unnoticed by central generating companies[2]. The ABT also faces opposition from central generating companies because the rewards for availability will be reduced compared to what they were under the previous tariff (based at that time on plant load factor). Asymmetry of incentives and disincentives is hard to justify if it is not related to periods of time. Ideally, it would be preferable to value availability according to system load, or to agree on a period for the maintenance of generation units.

---

1. For coal power plants, the ABT increases the target availability for full recovery of the fixed charge by 20 percentage points, from 65% to 85%.
2. NTPC appealed to the Delhi High Court against the CERC Order. The Court made an interim ruling that the previous tariff applied to NTPC would remain in effect after 1 April 2001, until payment arrears from SEBs to NTPC were cleared, and provided NTPC profits were all reinvested (CERC, Tariff for Power Stations from NTPC Power Stations from 1st April 2001 Petition 30/2001, 13.6.2001). NTPC declared that the "Availability Based Tariff (ABT) system ordered by the CERC will result in drastic reduction in its internal financial resource generation capacity. [NTPC] has estimated the changes in tariff principles and norms by the regulator, which are being applied from 1 April beginning with the Southern Region, will lead to a Rs 180 billion (USD 3.82 billion) cut in revenue generation over the period up to 2011-12. This will mean the addition of only 6,000 MW of new capacity instead of the planned 20,000 MW by 2012" (Financial Times, *Power in Asia*, 326, 17 April 2001).

Electricity Grid Code, Dec. 1999

The Electricity Grid Code was designed by the CERC to improve the quality of electricity transmission in India, one of the primary goals being to reduce the range of frequency and voltage fluctuations. The code covers interstate commercial exchanges using the ABT and lays down the rules, guidelines and standards for management of the power system in the most efficient and reliable manner. As of the beginning of 2001 the Code has not been fully implemented.

The CTU, POWERGRID, a commercial public entity, is responsible for the planning, maintenance and operation of interstate transmissions under the Electricity Law (amendment) Act, 1998. POWERGRID has the lead role in reviewing the Grid Code. Modifications to the Code must be approved by the CERC.

The Grid Code defines the role of the RLDCs, which have overall responsibility for real-time operation of the power system. Under the new code, RLDCs may overrule the SLDCs. The Grid Code also re-defines the respective roles of REBs and the CEA in planning and co-ordinating grid development.

The **planning code** for interstate transmission defines the relative responsibilities of the CEA and the CTU in planning the interstate transmission system, as well as the criteria for connection to the interstate transmission system.

The **operating code** for regional grids defines the rules to be followed by every regional grid, particularly on voltage and frequency control.

The **scheduling and dispatching code** clarifies how regional grids are to be operated as loose pools. This means that states have full autonomy over dispatching from their own generation units and their own captive power plants. This code sets up the procedures to be followed by Interstate Generating Stations (mega-power IPPs or central generation units) and state customers, for dispatch and use of power.

Box 2   Regulatory Actions to Improve Grid Discipline

A major disturbance in the Northern Grid took place on 2 January 2001, plunging all the states in the Northern Region into darkness for many hours[1]. The grid collapse was reported by the Northern Region Load Dispatch Centre (NRLDC) to the CERC, which investigated the incident.

Based on its investigation, on 15 January 2001, the CERC issued a first Order aimed at improving the functioning of the grid. The order attributed the grid collapse to the cumulative effect of various incidents. Most of them were linked to under-investment in transmission-control equipment, mismanagement and non-respect of

---

1.   Jammu and Kashmir, Himachal Pradesh, Punjab, Uttaranchal, Haryana, Rajasthan and Uttar Pradesh.

the Grid Code. The following specific problems were identified: non-availability of the Rihand-Dadri High Voltage Direct Current (HVDC) line; failure of converter transformers restricting capability of power flow; tripping of 220 KV and 400 KV lines and flash-over on the HVDC line. The CERC concluded that the main responsibility for the collapse lay with the state-owned Power Grid Corporation Ltd (PGCIL). The CERC decided not to impose penalties that would normally apply under Section 45 of the CERC's Order on the Grid Code, but warned that it would penalise those responsible for violating the code on future occasions. The CERC ordered:

- the PGCIL to streamline the functioning of RLDCs and to ensure that the states comply with the directives of RLDCs;

- the central government, the CTU (in this case PGCIL) and the state authorities to explore ways of improving the quality of existing equipment as soon as possible;

- the NRLDC to install loggers and voice recorders at all major control rooms, to set up a quick and simple procedure for time synchronisation, and to augment power transmission lines in consultation with the CEA;

- the National Power Corporation of India to follow the instructions given by the regional load dispatch centre (the report observes that the NTPC had been generating in excess of the schedule provided by RLDC, and that the NTPC was also less than fully compliant in reducing generation as it was instructed to do by the RLDC);

- the CTU and the concerned generators to implement the Grid Code. One of the Code's major provisions is the restoration of free governor action, that is the ability to adjust the stations' output to the grid needs automatically. Non-fulfilment of this provision more than one year after issuance of the Grid Code was a major failure;

- the Northern Region Electricity Board (NREB) to formulate an action plan for installation of under-frequency relays.

On 16 January 2001, the CERC issued a second order to enable the CTU to be able to face such a crisis in the future and restore power rapidly. The order identified the reasons for the inordinate delay in restoring power in the region: failures of governors on machines at Bhakra; non-availability of reactive support for the Dadri-Panipat transmission line; poor maintenance of air-blast circuit breakers at the Panipat sub-station; and failure of circuit breakers on the 132 KV Rihand-Singrauli line. The CERC directed the CTU (as well as the CEA) to review the existing procedures for early restoration of the grid in consultation with the parties concerned. Following these two orders, the PGCIL filed a petition for review, as it felt aggrieved by certain observations made by the CERC, but the CERC ruled that the arguments presented did "not call for any further review".

REGULATORY CHANGES AT THE STATE LEVEL

State Electricity Regulatory Commissions

SERCs are envisaged in the ERC Act, 1998, which gives state governments the possibility to establish them if they are considered appropriate (Section 17(1) of the Act).

The main functions of a SERC would be:

■ to determine tariffs for electricity (wholesale, bulk, grid and retail);

■ to determine the tariff for use of the transmission facilities;

■ to regulate power purchases and the procurement process for transmission and distribution utilities;

■ to promote competition, efficiency and economy in the electricity industry.

The new regulatory framework establishes a foundation for pricing electricity at the cost of delivering the service, thus paving the way for subsidy reforms. The ERC Act, 1998 stipulates that the tariff should be determined by the SERC in such a way as to reflect the cost of supply of electricity at an adequate and improving level of efficiency. It also stipulates that "the state commission, while determining the tariff under this Act, shall not show undue preference to any consumer of electricity, but may differentiate according to the consumer's load factor, power factor, total consumption of energy during any specified period or the time at which the supply is required, or the geographical position of consumers, the nature of supply and the purpose for which the supply is required." In case a state government determines that subsidies are required, the regulatory framework asks state governments to pay them, protecting the utilities' finances.

State Tariff Orders

Principles and Implementation

To date, several states, including Orissa, Haryana or Andhra Pradesh have enacted reform bills, outlining a restructuring plan for their power sector. Most of the Indian states have established a SERC and some have issued tariff orders (see Annex 3).

The most sensitive issue is that of retail tariff notification.

The ERC Act, 1998, contains general principles to guide the SERCs in setting tariffs:

■ a cost-plus methodology[1] can be used to ensure a minimum overall 3% rate of return to the seller;

■ cross-subsidies are allowed but should be limited to 50% of the cost of supply;

A pure cost-plus methodology for setting tariffs provides no incentive for reducing transmission losses. Determining the tariff requires identification by the distributors of the cost of the volume of energy that has to be generated or purchased to produce

---

1. That is a price determined as a fixed mark-up added to the observed unit costs.

a given volume of sales at a given amount of losses. In this framework, the larger the volume of energy produced or purchased for a given volume of sales, the higher the cost per kWh sold and the higher the average tariff required. To compensate for this, the SERCs seek to define an acceptable level of losses.

Critique

Because certain categories of consumers benefit from large direct subsidies, the Government of India has made the correction of price distortions a policy priority though the promotion of independent regulatory institutions at the state level.

Few SERCs have passed tariff orders. The state of Orissa is an often-cited exception. It anticipated the ERC Act, 1998, by passing a reform Act in 1995, and has passed three tariff orders since then (Sankar, 2000).

Implementation of the orders has failed to reduce transmission and distribution losses. In Orissa for example, where T&D losses are on the order of 40%, the initial objective of the OERC was to induce distributors to reduce them by six percentage points, allowing losses of only 34% (against 38% to 42.7% asked for by the distributors). Achieving this goal however, has proved difficult in spite of the relatively favourable conditions in Orissa, where sales to agriculture at subsidised prices account for hardly 4% of the total, and domestic sales account for 25%. In spite of the relatively small number of customers benefiting from subsidies, political interference in the implementation of price hikes has impeded the planned reduction of T&D losses. Disconnecting non-paying customer has been made difficult. But there has been some progress. A 19.35% tariff hike for domestic consumers has been allowed while high-tension tariffs have decreased, thereby reducing the cross-subsidies.

The challenge for distributors – private or public – is to reduce non-technical losses and to obtain satisfactory cost recovery for the energy they sell. The latter can only be done by increasing tariffs for agriculture and domestic consumers. The higher the value of sales, the more economically viable it will be to install meters and the more worthwhile it will be to make customers pay for their electricity consumption. The higher the price, the more sensitive customers will be to the quality of supply, and the less readily they will accept paying for those who are not metered or who steal electricity.

For these reasons, both the central and the state governments should:

■ allow tariffs to be based on development costs rather than historical costs. Price regulation based on historical costs does not allow distributors to recover amounts corresponding to improvement and development of the system. Tariffs should internalise development costs, especially investment in meters and in efficient accounting systems;

■ define and implement penalties for the distributors for inadequate supply quality. Penalties should depend on the frequency and duration of blackouts, brownouts and load-shedding, power frequency divergences and other indicators of the quality of the electric power delivered;

■ make electricity pilferage illegal and encourage innovative ways of distributing and selling electricity including the establishment and management of mini-grids by rural co-operatives.

Policy Reforms in States and Federal Support

Acknowledging the imperative for reforms at the state level, the Government of India decided in 2000 to develop individual contractual frameworks with states, conditioning its financial support on the implementation of reforms. The Memoranda of Understanding it signed with the different states affirm the joint commitment of the parties to reform the power sector, stipulate the reform measures that the state will implement and define the support that the Government of India will provide.

One of the explicit aims of these memoranda is to restore the commercial viability of the electricity sector, providing reliable and quality power at competitive prices to all consumers in the state. The main focus is on reforms in the distribution sector. Through measures such as privatisation of distribution, metering, and reducing pilferage, the Government of India is trying to improve revenue collection and to turn around the deteriorating ratio of average revenue to cost of supply. The Power Finance Corporation (PFC) is in charge of channelling the central funds to the states.

By January 2002, the Government of India had signed such memoranda with nineteen states. Several states have enacted reform bills for their power sector. Most of them have established SERCs (18 states by January 2002) and some of them have issued tariffs (11 states by January 2002).

PRIVATE PLAYERS' RESPONSE TO MARKET DEVELOPMENT

Independent Power Producers

Since October 1991, when private investment in the power sector had just been authorised, the Indian government has constantly sought to facilitate such investment.

As in many other countries, initial project solicitation was carried out in India through negotiation between state authorities and project promoters. Formal bidding procedures began in 1993 and spread more widely in 1995, as SEBs sought to gain better deals through transparency and competition. The bidding procedure could not really start until state governments had undertaken an integrated resource-planning exercise which enabled them to identify system needs, additional capacities required, technical and environmental characteristics and the mode of despatch. The next stage was the preparation of feasibility reports and obtaining different clearances and linkages for the projects (FSAs and PPAs). Most of these steps had to be carried out by the project promoters in an environment where few rules were clearly defined, let alone standardised. Since an IPP cannot trade power directly to a state other than the one in which it is set up, IPPs operate within the framework of a power purchase agreement linking them to a single state buyer. In other words, private power projects in India are not merchant plants like those found in fully liberalised markets. The Minister of Power wrote to all the Chief Ministers in October 1993, expressing the need to introduce competition by asking for price bids. The Ministry of Power issued guidelines for competitive bidding for private power projects in January 1995. Projects agreed by memoranda of understanding (MoU signed prior to 19 February 1995) were asked to use international competitive bidding to award engineering, procurement and construction contracts.

Private-sector entrepreneurs can set up companies either as licencees or as generating companies. A licencee holds a licence, issued by the state government concerned under Section 3 of the Indian Electricity Act, 1910, to supply and distribute energy in a specified area, which may or may not have a generating station. Generating companies can be privately owned. It is possible for a company to act as a generating company in one area and a licencee in another. The generating company can sell power to SEBs at a tariff based on the parameters applicable to generating companies.

Surplus electricity from auto-producers plants can be offered for sale to the SEBs. Guidelines have been issued to the state governments to facilitate early clearances of proposals and also to ensure effective measures such as wheeling surplus power from such plants.

A debt equity ratio up to four-to-one is permissible for all prospective private enterprise entrants, both licencees and generating companies. Up to 100% foreign equity participation is allowed.

A two-part tariff was announced on 30 March 1992 for the sale of electricity by private generating companies to state-owned utilities. ABT norms apply for thermal power projects growing out of international competitive bidding.

Customs duties are reduced for the import of power equipment and for machinery required for renovation and modernisation (R&M) of power plants. Since R&M is the cheapest and quickest way to add capacity, the central government decided to accord it highest priority. Renovation schemes involving capital expenditure of up to Rs 5 billion do not require the approval from the CEA.

Participation of the private sector in transmission projects is also encouraged.

Box 3   Private-sector Participation in Transmission Projects

India's power transmission and system operations are going through an extensive restructuring program in parallel with evolving state-level reforms. POWERGRID, India's central transmission utility, is the main implementing agency for this program. To encourage private investment in the transmission business, the central government enacted the Electricity Laws (Amendment) Act in August 1998, which gave transmission activity independent status and introduced the concept of central and state transmission utilities. While POWERGRID was announced as the CTU, the SEBs or their successors would be the state transmission utilities (STU).

Private-sector participation in transmission is limited to construction and maintenance of lines for operation under the supervision and control of CTU/STU. The private company will contract only with the CTU/STU for the entire use of the transmission line(s) constructed by the company and will be responsible for maintenance of the lines. Transmission charges payable to the company will be directly linked to the availability of lines. Guidelines for private-sector participation have been prepared.

Orissa has already unbundled energy production and transmission operations and completed the privatisation of distribution. Haryana, Andhra Pradesh and Uttar Pradesh are expected to follow suite. Gujarat, Kerala, Karnataka and Rajasthan have also taken initial steps toward privatisation.

POWERGRID plans a large number of transmission projects for which it needs private investment. These projects are estimated to cost around USD 2.62 billion and are to be carried out in association with the private sector through joint ventures or through the creation of independent power transmission companies (IPTC). In the joint ventures, POWERGRID will hold a stake of 26%. In July 2001, National Grid of UK and Tata Electric Corporation have been short-listed as joint venture partners.

Projects identified by POWERGRID for development through private investment

Name	Commissioning Year	Cost, billion Rs
Transmission system for Sipat-I (1980 MW)	2005	27.3
Transmission system for Rihand-II (1000 MW)	2006	11.4
Transmission system for Maithon Right Bank (1000 MW)	2006	6.2
Transmission system for Ennore (1800 MW)	2008	15.0
Transmission system for Karcham Wangtoo (1000 MW)	2009	6.2
Composite transmission system for Kahalgaon, north Karanpura-Barh (5280 MW)	2009	92.0
Transmission system for Neyveli thermal power station II & III (500 MW)	2009	12.5
Transmission System for Hirma-I (3960 MW)	2009	58.1

Sources: http://www.powergridindia.com / http://www.powermin.nic.in

Fast-Track Projects

As early as 1992, the Government of India identified eight large private-power generation projects and gave them fast-track status, with the primary objective of quickly bridging the demand-supply gap. Fast-track projects were supposed to receive guarantees from the central government that the developers would be paid if the SEBs defaulted. In 1994, the government decided to provide guarantees to the eight projects, but they were eventually issued for only two: Dabhol and Jegurudapu. In 1998, revised and limited guarantees were issued for three other projects.

Mega-Project Policy

First launched in 1995, the mega-project policy is intended to develop power plants with a capacity of 1,000 MW or more and supplying more than one SEB. POWERGRID is to develop a network to transmit power from these projects to other states. The mega- project policy concerns both the private and the public sectors. Among other benefits, private mega-projects enjoy customs exemptions for their imported equipment.

The Power Trading Corporation (PTC) was established to purchase power from the private[1] mega-projects and sell it to interested states. Payment was to be guaranteed through mandatory letters of credit and a right to funds from the beneficiary states if the SEBs failed to pay up.

---

1. Public mega-projects are expected to deal directly with SEBs.

Table 4 Fast-Track Projects

Project/ promoter	State	Capacity commissioned (MW)	Situation	Provisional cost (Rs. billion)	Fuel	Technology
Dabhol/ Enron	Maharashtra	Phase I: 740 (Phase II: 1444)	Fully commissioned (Phase II halted)	28 (+ 63)	Natural gas/ naphtha	CCGT
Jegurupadu/ GVK Reddy	Andhra Pradesh	216	Fully commissioned	8	Natural gas/ naphtha	CCGT
Godavari/ Spectrum Power Generation	Andhra Pradesh	208	Fully commissioned	7	Natural gas/ naphtha	CCGT
Ib Valley TPS/ AES Transpower, USA	Orissa	500	TEC obtained	24	Coal	n.a.
Neyveli/ST-CMS Electric Co.	Tamil Nadu	250	TEC obtained	12	Lignite	n.a.
Mangalore/ Mangalore Power Co.	Karnataka	1,013	TEC obtained	43	Coal	n.a.
Visakhapatnam Ashok Leyland and National Power Plc., UK	Andhra Pradesh	1,040	n.a.	n.a.	Coal	n.a.
Bhadravati/ Nippon Denro Ispat	Maharashtra	1,072	TEC obtained	46	Coal	n.a.
Total		6483				

Sources: GOI – Ministry of Power, Central Electricity Authority, Ministry of Finance.

The Chief Ministers' Conference revised guidelines for the Mega-Project Policy in December 1999. As a pre-condition for a mega-project, states are now required to have a regulatory commission. As of 2000, 18 mega-projects had been announced including four in the private sector.

Escrow Accounts

Escrow cover is one of the security mechanisms to mitigate commercial risk for the IPPs. Escrow was put in place in 1995 as a response to the limited ability of the Government of India to guarantee private-sector generation projects directly. Escrow accounts are opened by SEBs to allow IPPs to make claims for payment due. The cash flows (receivables) of the SEB from selected customers/distribution areas are deposited directly into the escrow account instead of being paid to the board.

Because distribution produces very low revenue, it is difficult for distributors to pay for the electricity generated by IPPs. Moreover, the states' public finances have deteriorated in the past decade. As a result, most of the states rapidly exhausted their escrow accounts. This system did not help in securing sizeable private investments for generation projects.

The Role of Multilateral Financial Institutions

Multilateral financial institutions are supportive of power-sector reform and of more general economic reforms aimed at mobilising investment and increasing economic efficiency.

In the early 1990s, the World Bank decided to finance mainly projects in states that "demonstrate a commitment to implement a comprehensive reform of their power sector, privatise distribution, and facilitate private participation in generation and environment reforms". This marks a change from the period before 1993, when the World Bank financed mostly large-scale generation projects. This strategy shift was justified by the fact that support to large generation projects had not contributed effectively to the emergence of a viable power system in India, as reflected in a subsequent report by the Operation Evaluation Department, *Meeting India's Energy Needs (1978-1999): A country sector review*. Accordingly, recent loans from the World Bank have gone to support the restructuring of SEBs (World Bank, 1999). In general, the loans are for rehabilitation and capacity increase of the transmission and distribution systems, and for improvements in metering the power systems in states that have agreed to reform their power sector.

The overall strategy of the Asian Development Bank (ADB) for the power sector is to support restructuring, especially the promotion of competition and private-sector participation. The ADB supports power generation projects if they are beyond the financial reach of the private sector, especially large hydroelectric projects. The ADB also supports rural electrification and small grids, especially when electricity service is not yet commercially viable. Like the World Bank, the ADB is providing loans for restructuring the power sector in the states and improving transmission and distribution. One of the latest ADB loans is also to support POWERGRID in integrating the Indian power system.

Box 4   World Bank and Asian Development Bank Support for Indian Power-sector Reforms

World Bank

▪ **Orissa** was the first state to launch a major overhaul of its power sector. To support the state's program, the Bank provided a USD 350 million loan in 1996.

▪ A new lending instrument, the Adaptable Program Loan (APL), has been developed and is now the cornerstone of the World Bank's approach to supporting India's state power reforms. **Haryana** was the first state to benefit from the new approach. In January 1998, the Bank approved a USD 60 million APL to support the first phase of Haryana's program to restructure its ailing power sector. The loan is the first of a series of APLs totalling USD 600 million that the bank plans to provide over the next eight to ten years to support the program.

▪ The approval of a USD 210 million APL for **Andhra Pradesh** followed in February 1999, the first in a series of APLs totalling up to USD 1 billion that the Bank plans to provide over the next eight years. The World Bank is supporting Uttar Pradesh through a USD 150 million loan approved in April 2000. These loans will help to transform the state's power sector from a major drain on the state's budget into a source of revenue for priority sectors.

■ **Rajasthan** is receiving similar assistance under the Rajasthan Power Restructuring Project, and it is expected that other states will be inspired to undertake meaningful power-sector reforms.

The Bank has been closely involved in POWERGRID's development from the beginning, funding its co-ordination and control facilities, transmission lines and substations, and various institutional development activities. The process of establishing POWERGRID involved transferring the transmission components of several earlier Bank loans and IDA credits. Along with the USD 350 million POWERGRID System Development Project in 1993, these transfers brought the Bank's total investment in POWERGRID to about USD 1.5 billion. A loan for a follow-up project, POWERGRID II, has been processed.

Asian Development Bank

In October 2000, the Asian Development Bank approved a loan of USD 250 million for India's power sector. This money will be used to establish a national grid for interstate power transmission. The ADB also extended its partial credit guarantee for raising another USD 120 million from commercial banks. On 13 December 2000, the Asian Development Bank approved two loans totalling USD 350 million for the power sector in the western state of Gujarat.

The multilateral financial institutions are emphasizing integration of national networks. In India that means projects that foster integration of the five regional power systems, as well as integration of South Asian power systems. These include high-voltage transmission lines and institutional frameworks and investments that favour bulk-power trade between power systems.

The financing institutions have long understood that the way to reduce the supply-demand gap is to improve existing systems by reducing losses, improving revenue collection and market development. To do these things, grid integration is a necessity. With reduced loans available for energy projects, the current need is to maximise the yield of investment in the sector.

Box 5   Dabhol Power Project

The Dabhol power project of the Dabhol Power Company (DPC) has long been a key independent power producer's project and an important part of India's efforts to attract investment in electricity generation. Well before its majority stakeholder, ENRON, filed for bankrupcy at the end of 2001, the Indian project was stopped following default of payment by the electricity buyer and DPC's contract termination at the begining of 2001.

DPC began producing electricity in May 1999, selling it to the Maharashtra State Electricity Board (MSEB). It is located on the West Coast of India, approximately 170 km south of Mumbai in the state of Maharashtra. In 1993, the Indian government gave fast-track status to this project. ENRON took a 50% stake in DPC. In its final phase, the Dabhol power project is planned to have 2,184 MW of combined-cycle units using natural gas. In the first phase, a 740 MW combined cycle capacity unit fuelled by naphtha was commissioned in May 1999. The second and final phase, a 1,444 MW unit was initially scheduled for commissioning in the fourth quarter of 2001. Its construction has been stalled following the beginning of the problem between DPC and the MSEB.

The natural gas will be provided via a 5-million-ton-capacity LNG terminal and re-gasification plant located next to the power plant and built along with it. Contracts for 3.7 million tons of LNG had been signed by ENRON with Gulf exporters. If it is finalised following the initial plans, the Dabhol project could eventually represent around 2% of Indian power generation capacity. Despite its size, it was not considered a mega-power project, because it supplies power to only one state. So it did not benefit from financial support by the Government of India. However, because it is a fast-track project, it obtained a guarantee from the Government of India.

Initially, the project suffered many delays because of disagreements between the MSEB and DPC, particularly over the power purchase agreement. The PPA was re-negotiated in 1995, before the first unit began operation. In 2000-2001, the average tariff charged by DPC for Phase I was 4.8 rupees/kWh, much higher than the average cost of power purchased by MSEB (Rs 2.2/kWh). This high price was explained by increases in naphtha prices on the international market and the plant load factor decrease due to a lower use of the plant for merit-order reasons, as required by the states's electricity regulatory commission. It is also explained by the high capital cost of the plant[1]. Since the end of 2000, MSEB has regularly defaulted on its payments. MSEB alleged that DPC failed to supply power within three hours of demand from cold start in January 2001. On 7 April 2001, DPC claimed it could not fulfil its contractual obligations due to politically inspired circumstances beyond its control.

Judging who is at fault is beyond the scope of this report. Our aim here is to evaluate general policy implications linked to the DPC experience. The high cost of power generated by DPC seems to arise mainly from large capacity and relatively small demand in the state and poor administration by public authorities.

In 1993, when asked by the Government of India to review the project for possible financing, the World Bank pointed out that "the project was not the least-cost choice for baseload power generation compared to Indian coal and local gas. Even if local fuels are not available, imported coal would be the least-cost option". When dispatching the units of a generation power system according to merit order, a

---

1. The overall cost of the project is USD 2.9 billion. If the USD 700 million cost of the re-gasification plant is deducted, the unit cost is around USD 1,000/kW, higher than the international standard of USD 500-700/kW for combined cycle units.

dispatcher will first choose those with the lowest variable costs. After some hydroelectric units, these would be nuclear and coal units. In general, and despite their high efficiency, combined gas-cycle units would be ranked after most coal-fired units because of the high price of natural gas compared to coal. If the existing capacity of coal-fired units is large enough, it is highly probable that new natural gas-fired power plants would be used only as intermediate-load units (with load factors typically ranging from 20 to 50%). Once low-variable-cost coal units are retired, and assuming that little new coal capacity is introduced, natural gas units would move up in rank and be used for baseload. The entire natural gas infrastructure has to be built from scratch. Demand for power has to be large enough to make the new natural gas supply chain economically viable. Investigations carried out by the Energy Review Committee led by Mr. Godbole[1] show that actual demand for power was much lower than optimistic initial projections. The DPC experience points to inadequate oversight and preparation by the public institutions involved in approving the project. It also demonstrates lack of due diligence by some of the private players.

Lessons and recommendations

Several lessons can be drawn from the DPC experience:

DPC Phase 1 is the largest independent power producer to have begun operation in India since 1991. DPC is a symbol of the government's ability to conduct business with foreign direct investors. A solution to the present deadlock is essential;

contracts need to be respected. Re-negotiation could be sought only if and when the economic environment changes so radically that the existing contractual conditions run seriously counter to the public interest;

oversight by public institutions involved in approving projects needs to be improved, as does their understanding of market dynamics;

development of generation capacity should be better monitored at central government level, or at least at regional level, using least-cost as a basic criterion;

creation of a regional bulk-power market ought to receive more priority, along with measures to facilitate interstate bulk-power exchanges;

for energy security and environmental reasons, the development of natural gas infrastructure should be considered by the Government of India in order to diversify the power mix. In order for natural gas, especially LNG, to be cost-competitive, however, the infrastructure should be developed on a regional or national rather than a state basis.

---

1. At the request of the Government of Maharashtra after DPC's termination notice.

Captive-power

In India, captive-power, or auto-production capacity almost doubled in the 1990s and now represents around 15,000 MW[1]. Typical generation cost for these captive power plants is 2.5 rupees/kWh compared with an average 1999-2000 SEB tariff for industry of 3.5 rupees/kWh, and around 1.7 rupees/kWh for sales by the National Thermal Power Corporation to SEBs.

The central government has issued guidelines to promote auto-production projects. However, implementation of these guidelines at the state level has not been very satisfactory. States often try to maintain monopoly conditions for the SEBs through measures that are not favourable to auto-production, such as high wheeling and banking charges; high charges for remaining connected to the grid; low purchase rates for surplus power and non-permission for third party sales.

Critique

Since 1991, many developers have proposed generation projects. More than two hundred projects have been the subject of memoranda of understanding between developers and state administrations. Of the 57 projects which have received technico-economic clearance from the CEA, 47 have foreign equity participation.

Table 5

Status of Private Power Projects, as of March 2001

Description	Number	Capacity (MW)
Projects techno-economically cleared by CEA		
• Thermal	52	27,860
• Hydro	5	1,516
• Total	57	29,376
Detailed project reports under examination in CEA		
• Thermal	8	2,554
• Hydro	1	70
• Total	9	2,624
Private power projects which have been commissioned	25*	5,370
Private power projects under construction	17*	5,149

* This includes projects which do not require the techno-economic clearance of CEA and licensees.
Source: Ministry of Power.

Since 1991 however, the total additional installed capacity from the private sector has remained limited. Most of the targets for additional installed capacity, in the Government of India's Eighth five-year Plan and those for the initial years of the Ninth Plan have not been met. Capacity increase throughout the 1990s has fallen far below the target of 40% more generation capacity called for by the government. Between 1992 and 1998 (Eighth Plan), total private sector capacity was 1,264 MW versus a target of 2,810 MW. The situation improved in 1999-2000, and the target was slightly surpassed although it should be noted that the target was considerably lower than in previous years (see Figure 11). In January 2002, the government adjusted its target and targeted 20% of future additional capacity to come from the private sector, against 40% before.

---

1. Estimates of auto-production capacity vary widely between the central government and the specialised press. The estimate given is on the conservative side; it may not take into account units under 1 MW that together may represent another 7,000 MW.

Figure 11   Additions to Installed Generating Capacity, MW

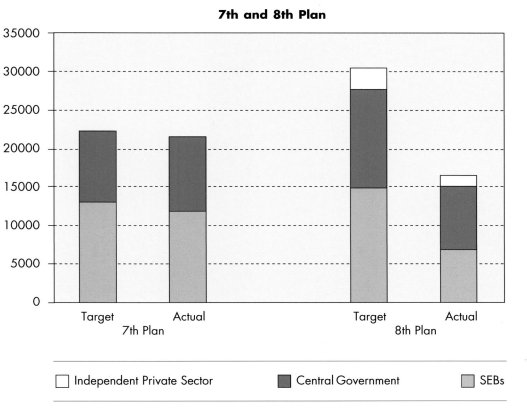

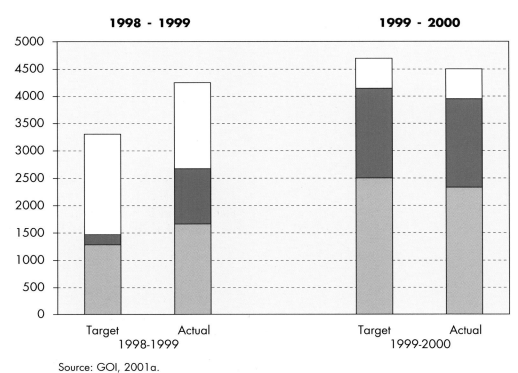

Source: GOI, 2001a.

Even though private-sector generation increased in the second half of the 1990s, private-sector generation still remains limited.

Figure 12   Electricity Generation per Category of Market Player, Million kWh

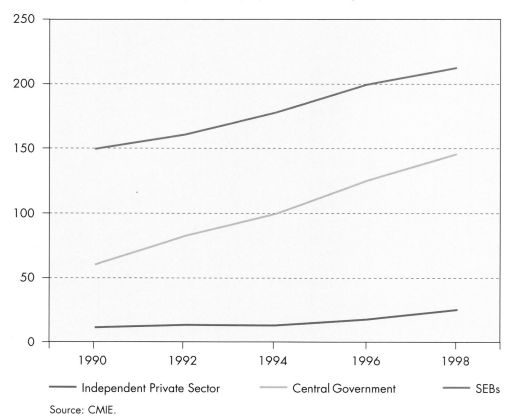

Independent Private Sector   Central Government   SEBs

Source: CMIE.

Several of the foreign companies that initially invested in the Indian market withdrew from it, often citing the low returns on their investments in India compared to markets elsewhere.

The Government of India gives the following reasons for the slow progress of private-sector investments in the power sector, especially in generation:

■ delays in achieving financial closure are mainly due to the poor and deteriorating financial health of the SEBs, who do not have the financial capability to support more than a few projects;

■ states are unable to sustain sufficient escrow accounts. This is required by almost all the institutional foreign investors financing IPPs. States have had difficulties determining the amounts available for escrows. In several cases, the amounts identified by the state governments as being available have not been accepted by the financial institutions. The inability of the state governments to provide escrows for all IPPs has also led to litigation by some IPPs;

■ there have been delays in finalising essential contracts such as PPAs, FSAs and FTAs[1];

■ there have been court cases in the form of Public Interest Petitions.

With the reforms initiated by several states, the Government of India expects that the financial condition of their utilities will improve, paving the way for greater foreign investment. Most of the generation projects that emerged from the private-sector investment policy initiated in 1991 were for captive uses. The average size of IPPs is around 300 MW, which limits the possibility of economies of scale. Only three combined cycle power plants, at Dhabol, Hazira and Paguthan, exceed 500 MW. None of them is fuelled by coal or lignite. Although coal is a cheap and abundant resource in India, guaranteeing low cost energy, few of the IPPs choose coal as a fuel. The primary reasons expressed by investors for not using coal are difficulties in completing fuel supply agreements with the state-owned coal industry and in securing railway transport for coal. A second reason is the poor and variable quality of coal that leads to forced outages and makes it difficult to guarantee availability to the purchaser. The combined cycle power plant at Dabhol is the only sizeable IPP to have contracted with a SEB without having part of its revenue guaranteed directly by industrial customers. For the time being, natural gas combined cycle IPPs seem to be the most promising, but the recent experience of Dabhol illustrates that some outstanding issues remain. Most of the projects that have succeeded are of small size, use natural gas as a fuel and are located in credit-worthy states.

Auto-production is a short-run solution that the states should certainly not overlook. States should facilitate its development and its connection to the grid. Otherwise, if auto-production remains mostly a stand-alone solution, there could be economic loss in the long run where stand-alone systems do not benefit from the advantages of the grid: the ability to move power to different users according to their needs and the ability to reduce the capital costs required to satisfy a given demand. An analysis of this revenue loss could be a real eye-opener.

Recent experience in implementing policies to promote private investment indicates two more serious drawbacks:

■ all steps in project clearance need to be streamlined to improve transparency, efficiency and oversight. The number of administrations and institutions involved should be reduced so that responsibilities are not diluted (or even partially hidden), at both the state and central level. The number of political entities involved in the design of essential contracts such as FTAs, FSAs and PPAs should also be reduced, especially at the central government level;

■ there are often differences between the Government of India and the states about implementing the central government's reform agenda. The validity of giving the

---

1. The steps necessary to ensure the development of a project in India are complex and lengthy. Investors are required to update their expertise regularly and to monitor and adapt to numerous procedural and regulatory changes from the beginning of a project to the start of production.

states great responsibility for the future of the power sector is questionable, in a situation where markets need to be expanded beyond state borders. Recent efforts by the central government to gain a consensus on reforms, as demonstrated by the memoranda of understanding signed between states and the Government of India in 2001, are steps in the right direction, but final responsibility still lies with the state governments.

IV. REMAINING CHALLENGES

The current financial difficulties of the public power sector at state level are largely due to lack of political commitment, poor organisation and inadequate oversight. Improving the financial health of the sector is a crucial short-run issue. Improving the performance of the distribution sector by increasing revenue is probably the most urgent issue.

Improving the financial health of the sector also means avoiding the temptation to burden market players with social responsibilities. Nevertheless, the need for market access for the large share of the population that cannot afford electricity cannot be ignored.

There are also longer-run issues affecting the sustainability of the Indian power system. One of these is integration. Primary energy diversification, costly infrastructure development for gas, rapidly increasing demand for electricity are all factors calling for a larger market, better co-ordination between states and ways to improve resource allocation and exploit economies of scale at the national level.

The fourth challenge is the power mix. Policy-makers need to make the best use of India's existing primary resource endowment, mainly coal, and hydro power.

While we focus on these four challenges in the belief that they have not received enough consideration, other issues could have been addressed as well. One that immediately comes to the mind is the environmental impact of power-sector development. The main domestic energy source used for electricity is coal. It is of low quality, with an average ash content of 40%. Because of the need to boost electricity supply, environmental concerns have not been high on the Indian agenda for power reforms. Efficiency is likely to improve with market development. But the expected substantial increase in the use of coal by the power sector will still pose a crucial challenge to sustainable development because of a serious increase in polluting emissions. The Government of India will need to take more active steps to mitigate emissions from the power sector in the future. Current emissions from the power sector and the outlook for future emissions, as well as detailed policy analysis and recommendations on the environment and the Indian power sector can be found in IEA, 2000a, World Bank, 1999c and Wu & al., 1998.

THE POOR PERFORMANCE OF DISTRIBUTORS

A General Assessment

Distribution and supply, unbundled from generation and transmission activities, will probably remain the responsibility of the states and increasingly be controlled by state Electricity Regulatory Commissions. Current reforms focus more than in the past on the distribution sector, which in most states is still controlled by SEBs.

The performance of a distributor of electricity can generally be assessed using two criteria: good-quality and reliable electricity supply; and supply of electric power at the least cost.

Because those objectives are interdependent, reform policies in the distribution sector must address them together. A clear assessment of the situation is required at the outset. Parameters include: the quality of the power delivered, the quantities sold, consumption patterns and load charges, as well as the actual financial situation of the distributor (be it an unbundled public entity or a department within a SEB). It is essential also to analyse the record of existing private companies involved in distribution (such as BSES and AEC).

The Current Situation of the Distribution Sector

The sector is hard to assess since the quality of available statistics on sales and consumption is inadequate, and the statistics often contradict each other. Poor information about electric power consumption makes it difficult to design and operate a distribution system and discourages distributors from managing their systems in a cost-efficient manner. Investments in distribution cannot be allocated rationally, and so the reliability of power supply is reduced.

Electricity is often under-priced, hampering the financial viability of the sector. The most appropriate allocation of electricity is obtained by cost-reflective pricing, but for historical, political or specific policy reasons often linked to social goals, pricing in India may not be cost-reflective. When the gap between the actual and the cost-reflective tariff is too large, the job of the supplier is very difficult. Metering costs too much. Equipment becomes overloaded and deteriorates. Maintenance and investment operations are delayed or even cancelled, and a good-quality supply of electric power can no longer be ensured. The larger and more affluent consumers increasingly resort to individual investment to ensure an adequate supply of electricity in the form of batteries, inverters, stand-alone diesel or petrol generators.

Box 6   Improving Cash Collection in the Retail Electricity Sector: the Russian Example

Low cash recovery from 1995 to 1999 was one of the main factors hampering the Russian electricity sector. Unified Electricity System Ltd (UES) began cutting off non-paying customers in July 2000. The company reports that during the year 2000, settlements came to 105% of charges, showing some recovery of past arrears. The share of cash payments increased from 35% in 1999 to 83% in 2000. Starting 1 January 2001 all non-cash settlements were prohibited. UES announced that cash payments during the first quarter of 2001 represented 92% of the total due, a promising sign because collections even held up in winter, when the heating component in total payments is high.

Since many of the largest non-payers are Russian state bodies, a major part of the solution to the non-payment problem lies in the hands of the federal and local governments. It was therefore an important step when the federal government increased the budgets of administrations in 2000 to allow them to pay their energy bills.

In 2000 UES used various methods to improve collections, including disconnection of non-payers, establishment of a system of prepayments and the use of letters of credit, introduction of limits on deliveries to state organisations, and tightening of controls over electricity consumption by public-sector consumers.

UES had little choice but to crack down on non-paying customers, when Gazprom – the largest Russian gas supplier – demanded 100% payment from UES for gas deliveries in the first quarter of 2000. This heightened tensions between Gazprom and UES, although gas deliveries were maintained with only slight reductions (compared to the cuts threatened by Gazprom). But reduced gas deliveries obliged UES to resort to more expensive heavy fuel oil and coal to make up the difference. The hard budget constraints imposed on UES by Gazprom forced UES to impose those same constraints on its customers. The results have been impressive.

Source: IEA, 2002.

Non-payment is the main issue faced by the Indian power sector. In certain areas such as New Delhi, unpaid consumption (billed or not) amounts to one half of generation. This problem is exacerbated by the absence of decentralised, cost-based decision-making by the numerous technical staff involved in distribution and supply operations.

This deteriorating situation has implications upstream. SEBs are unable to pay CSUs[1] on time. In 2000-2001, for example, the unpaid dues of all SEBs to the National Thermal Power Corporation represented 80% of NTPC's turnover.

Continuing Reforms

Privatisation of Distribution

Calls have often been made, particularly by multilateral financial institutions, to unbundle the power sector, and to privatise distribution under the new electricity regulatory regime. These calls are the result of assessments showing that large vertically-integrated public utilities often have difficulties in managing distribution and sales.

Although reluctant at first, most state governments are gradually implementing reforms based on unbundling transmission and generation from distribution. This trend needs to be encouraged. However, many issues remain. Should distribution activity be corporatised or privatised? What is the optimum size of a distribution franchise? What are the various roles of governments and regulatory commissions in regulating distribution activity? Answers to these questions vary from state to state.

Whereas nearly all SEBs show very poor distribution performance, private companies involved in distribution for a long time such as BSES in Bombay, AEC in Ahmedabad and SEC in Surat are performing relatively better and earning profits. On the other hand, the experience of the new distribution companies in Orissa, CESCO in particular,

---

1. In 2001, after a report recommended transforming power arrears into government bonds, the Government of India launched negotiations with the states for this purpose.

Table 6   Arrears Owed by States to CSUs, as of 31 March 2001, Billion Rupees

STATES		CSU	
Andhra Pradesh	6.3	REC	−34.7
Arunachal Pradesh	0.3	NTPC	−153.9
Assam	11.4	NEEPCO	−10.1
Bihar/BSEB	60.6	DVC	−27.9
Gujarat	9	NHPC	−32.8
Goa	0	NPC	−22.9
Haryana	12.5	PFC	−0.2
Himachal Pradesh	1.3	PCIL	−9.9
Jammu & Kashmir	12.1	Coal Companies	−65.3
Karnataka	7.4		
Kerala	7		
Madhya Pradesh	47.1		
Maharashtra	14.3		
Manipur	2		
Meghalaya	0.6		
Mizoram	0.5		
Nagaland	0.9		
Orissa GRIDCO	5.7		
Punjab	6.5		
Pondicherry	0.7		
Rajasthan	6.8		
Sikkim	0.5		
Tamil Nadu	13		
Tripura	0.6		
Uttar Pradesh	52.8		
WBSEB	31.1		
WBPDC	6.9		
DPL	2		
Delhi (DVB)	37.8		
Total	**357.7**	**Total**	**−357.7**

Source: GOI, 2001a.

shows that transforming state-owned distribution franchises into economically viable private ones is not an easy task. In July 2001, AES, the US-based company controlling CESCO in Orissa, stated that: *"If satisfactory resolution of these issues is not expeditiously reached, AES will be forced to abandon its commitment to the distribution company"*. In January 2002, AES decided to withdraw from India and sell the Indian company to its employees. Issues mentioned, among others, was the need for tariff increases, multiyear tariffs and enforcement of law and order (against electricity pilferage, and meter tampering).

In correcting a situation with high T&D losses, there is always a transition period during which the system has to be cleared up, reducing losses without excessively burdening distributors and consumers. When a large share of distributed electricity goes unpaid, it is difficult for distributors to collect their bills or to pass on the costs of non-paying customers to paying customers. In other words, the main issue is who should pay – the paying customers or the distribution franchiser? There is no single answer. The least we can say is that it is the government's responsibility to take legal action against electricity pilferage. It is also the government's responsibility to help find ways to cover a share of the costs associated with the transition.

If privatisation can be achieved while ensuring a level playing field for all private players, Indian and foreigner, it will certainly help introduce cost-efficient management of distribution and supply. Private distributors will point out the true problems of distribution. They will separate technical losses from non-technical ones, to address the latter.

But privatisation will not make distribution a profitable sector overnight. There is ample room for the government to initiate measures to unbundle and corporatise the existing distribution activities of SEBs and to improve their control. Because income structure and consumption patterns in India are idiosyncratic, the structure and institutional framework of the distribution sector may not be an exact copy of what exists in developed countries. Innovation should be encouraged. Delegating responsibility for low-voltage distribution to groups of consumers – through co-operatives, for example – is a possible avenue to explore. Even with publicly-owned companies, the creation of decentralised profit centres – with 10,000 to 50,000 subscribers and staff empowered with commercial responsibilities – is badly needed. This would be a preliminary step. Economies of scale would be affected at a later stage by concentrating distribution once the system is operating smoothly.

The Choice of Tariff Methodology

So long as the market is not fully competitive, prices have to be regulated. The issuing of tariff orders is an important activity for a State Electricity Regulatory Commission. It raises the issue of tariff methodology and ways to adjust the revenue of the distribution utilities.

So far, tariffs have been established using cost-plus methodology. Now, the SERCs can set tariffs based on the cost of service, tariffs based on performance, or tariffs that fluctuate along with an external reference, like the retail price index. In most countries, all three methods are used. In India, Performance Based Ratemaking seems the most appropriate approach (Ahluwalia, 2000) since the main goal is to reduce losses and improve the distributors' performance. Sustained – though moderate – inflation in India tends to justify using some element of price indexation in the tariff.

Tariffs should take into account only the load profile and geographical location of the consumer in determining when and what to charge. Industry should therefore pay less per unit than households, since consumption by industry is far more geographically concentrated and often has a flatter load than that of households. Exceptions for social reasons can however be made, as discussed below.

ELECTRICITY PRICING AND MARKET ACCESS

Implementing more cost-reflective electricity prices when a large part of the population lacks access to electricity is a difficult task. Reforms need to benefit a large majority of the population in order to be accepted. The fact that about one-third of India's population lives below the poverty line raises the question of the initial cost of providing access to energy markets[1]. Hence, as in many other developing countries "subsidies are likely to remain a key part of pro-poor energy policies for some time. Traditional ways to deliver subsidies often fail to help the poor. The challenge for governments is to find better ways of delivering subsidies" (ESMAP, 2000).

Healthy competition in the energy sector will eventually bring down costs and assist capital formation. In the short run, however, the consumer price will increase to reflect the costs of providing energy service without subsidies. In a country like India, there would be two negative effects of pricing energy at cost. Some consumers who are not able to pay the price may lose access to commercial energy. In the longer run, additional investments would concentrate on profitable market segments, limiting access for the part of the population unable to pay[2].

The benefits of society from the access to energy are easy to identify: improvements in health and hygiene through refrigeration and water heating, the immeasurable advantages of electric lighting and increase in workers' productivity, to name just a few. Another external benefit from widespread access to energy is the narrowing of social gaps But these external benefits are difficult to quantify and cannot be expected to be accounted for by energy markets. In a completely free-market system, energy is likely to be under-provided. In this respect, initial access to energy services qualifies as a social good.

Energy pricing should aim at cost recovery. But in a situation where energy is likely to be under-provided by the market, the need for some form of support to promote access to the energy service for poor households is acknowledged (ESMAP, 2000; WEC, 2000). However, the budgetary costs of subsidies are high (IEA, 1999). Financial support needs to be targeted and limited by time or income to avoid the regressive effects of subsidising energy users who are able to pay. The cost may be paid through a system of cross-subsidisation, or directly by the state, if public finances permit. The latter is preferable. The cost of a subsidy ought to be clearly identifiable; the costs of cross-subsidies might be hard to identify.

A subsidy scheme should have minimal implementation costs. In this respect, a system providing the subsidy at the supply level, such as in India has one advantage: its administrative costs will probably be lower than a system providing financial support directly to the demand-side through vouchers or income-support schemes. Direct support for demand, with the subsidy going to the consumer to cover the costs of his connection, and/or consumption, would be more efficient and cost-effective, but more

---

1. In 1993-94 35% of the Indian population lived below the poverty line. The poverty line is defined as the monetary equivalent of a minimum daily calorie intake (2,400 calories per person in rural areas and 2,100 calories per person in urban areas).
2. A similar situation would arise when the energy service is to be provided to consumers able to pay the price at the cost of their consumption but unable to afford the additional cost incurred by an extension of the central grid or the energy network. In this case, however, the access issue may not need to be solved by direct financial support, since decentralised energy technologies might offer a profitable alternative to grid-connected or centralised energy.

expensive to administer. It would require clear rules to identify and transfer the financial support to the beneficiaries. Various options exist for excluding unwanted beneficiaries. Each of them has different budget implications.

Apart from a general under-pricing of electricity as practised in India, *ad hoc* subsidy schemes have been established by private investors and public utilities in developing countries at the request of local administrations or in response to populist political pressures. For example, progressive electricity tariffs (also called social tariffs) are applied to the household sector in India, and in several other developing countries: Cambodia, Vietnam, Ivory Coast, South Africa, Costa Rica, Gabon. The principle is to charge larger consumers more than smaller ones. In most cases, a system of cross-subsidies between consuming categories allows the utility to recover the cost of delivering the electricity service. Very often, however, electricity prices are too low to recover costs and an additional system of cross-subsidies is necessary. This burdens industrial or commercial consumers with subsidies for the household sector. In 1998, the Indian central government launched the Kutir Jyothi Yojna programme (literally "light for small houses"). Under this programme, SEBs must connect households under the poverty line. The government and the SEBs provide grants up to a maximum of 1,000 rupees per connection with the installation of a meter or 800 rupees per connection without a meter. Implementation of the programme has been hindered by difficulties in identifying eligible households and by the SEBs' severe financial problems.

Such schemes need to be rationalised to reach their social target without hampering the efficiency of the whole system. The mechanism should not impose a financial burden on the utility providing the support. The threshold should not be so high as to encourage consumers to remain in the lower-consumption category, nor should the progressive tariff beyond the threshold be too steep.

The main challenge is to anticipate properly the overall cost of the subsidy mechanism. An attempt is made below to estimate the direct cost of demand-side support to electricity access.

As much as possible, access of the poor to electricity must be addressed by instruments of social policy, not by electricity pricing. If the poor cannot pay the full costs, then the difference to full costs has to be paid out of the state budget.

Box 7   The Cost of Subsidising Low-income Consumers' Access to Electricity

Let us assume that the government decides to facilitate the access of poor households to electricity by providing them with a minimum requirement, that is a lifeline system, where a financial support covers the consumption of a fixed monthly quantity of power, as well as the expenditure for their connection to the grid. Let us also assume that low-income households in Indian urban areas consume roughly 50 kWh per year[1]. What would be the total cost of such a system?

---

1. Poor households consume small quantities of electricity. A field survey made in a large city of south India in 1994 (Alam & al., 1998) showed that electricity represents one-fourth of the total energy consumed in the household sector (the rest is fuelwood, kerosene and LPG). In that survey, the lowest income groups consumed an average of 7 kWh per capita and per month (57 kWh per household). The figure for the richest income group was 41 kWh per capita (180 kWh per household) and the average was 15 kWh (90 kWh per household). The choice of 50 kWh as a threshold is debatable and probably on the high side. Similar progressive tariffs in other developing countries have sometimes supported lower consumption levels (e.g. up to 20 kWh per household per month).

The expenditure required to give the targeted population access to minimum electricity service has two components:

- the connection cost either through the central grid, or through local grids based on decentralised electricity production [C];

- the cost of poor households' daily consumption of power [E].

Note that the category "poor households" here refers to the part of the population that will benefit from easier access to power and does not necessarily refer to households with income below the poverty line.

The first component is a non-recurrent expenditure. The second is recurrent. The first component is significant in a developing country where the need to connect domestic customers is great and where the connection expenditure may be significant compared to the economic value of the electric supplied.

The financial transfers involved in the subsidies can take various forms. The mechanism with minimal administrative costs might be preferred. For example, money could be provided directly to the service provider, or to the final consumer, through a fixed amount deducted from the electricity bill.

The total cost of the subsidy will be [C] + [E] where:

[C] = (C * p) / H and

[E] = (S * P) / [H * (K*(1-B))]

with:

C = average connection cost (per household)

p = number of poor urban population to be connected

H = number of persons per household

S = marginal supply cost of power for residential consumption (production + T&D)

P = estimated poor population

K = chosen lifeline consumption level

B = fixed chosen percentage of electricity billed and paid for

In the present case, the calculation of [E] is based on a simplified lifeline rate system. All consumers are assumed to be billed for their electricity at marginal cost except consumers with consumption below the chosen lifeline level. The latter are charged a fixed proportion of the actual marginal production cost of the electricity service. This provision facilitates management of the financial transfer to households as it can be handled directly by the electricity provider. At the same time, it avoids supplying a totally free service which could give the wrong signal to consumers.

The supply cost of power for household consumption is estimated at 3.4 rupees per kWh[1]. Moreover, we assume that one third of the existing customers and half of the additional households to be connected each year will benefit from this lifeline rate. We also assume that the price charged to this category of the population will be one rupee per kWh, and that four million households will be connected each year[2].

On these assumptions, the maximum annual direct expenditure to be borne by the economy would be 44.7 billion rupees (roughly USD 1.1 billion): 11 billion rupees for connection charges for new customers and 28 billion rupees for the consumption of poor households. This is likely a maximum as the subsidy volume is calculated on the assumption that subsidised households consume their entire 50 kWh per month. Actual average consumption would probably be much lower. Special attention would have to be paid to the regular increase over time of the total direct expenditure as a result of additional consumers coming in (assuming the share of poor customers remains constant), if the support mechanism is maintained.

As an indication, the direct expenditure or cost of this support mechanism would be at least four to five times less than the current cost of subsidies for electricity consumption which amounts to 187 billion rupees, or USD 4.5 billion[3].

Figures used in the calculation (1997 data)

Indian population (millions)	980
Number of households (millions)	163
Households living in electrified zone (%)	90
Number of households living in electrified zones (millions)	147
Domestic customers (millions)	70
Domestic customers in the total number of households (%)	43
Overall domestic consumption (TWh)	59
Average annual observed consumption (kWh)	846
Distribution lines 500 kV and under (km)	3,108,830
Meter cost: purchase + installation for 1 Phase Electromagnetic kWh in rupees	583
Connection cost per customer (rupees)	2,783
Assumptions:	
Number of persons per household	6
Length of line to be installed per new customer (m)	20

Sources: CEA, 1998a; CMIE, 2001; IEA, 1999; RSEB, 1999 and IEA calculation.

---

1. IEA, 1999.
2. The current rate of connection is slightly above three million.
3. As estimated in IEA, 1999. This is a conservative figure as it only accounts for subsidy transfers to households and industry.

INTEGRATION OF THE INDIAN ENERGY SECTOR

What Integration?

In OECD countries, the system of centralised generation is gradually being challenged by distributed generation. And electricity markets are rapidly becoming integrated regionally, straddling national borders to exploit economies of scale. Integration is legitimate to reduce the cost of producing electricity and to boost production capacity.

In the electricity-supply industry, integration generally means the development of an interconnected system operated by a single entity, either an independent system operator or a central dispatcher. The feasibility of integration depends on the central institution's ability to harmonise state regulations, a difficult task given possible political opposition.

In India, integration is crucial for two reasons:

■ to encourage investment. In this case integration implies certain clear policy objectives such as electricity access for all, energy subsidies, energy security, and social and environmental goals. But it also means streamlining control of the power sector and facilitating co-operation with other energy sectors;

■ to take advantage of a larger market's scale effects, to reduce the overall cost of electric supply and to facilitate exploitation of additional primary energy sources. India has not yet reaped all the full benefits of a large centralised energy infrastructure. India has to build additional combined-cycle gas turbine plants and exploit its rich hydroelectricity potential. To do so, electric power demand should be pooled to mitigate the commercial risks for private investors. Recent experience shows that even rich states, such as Maharashtra, which are willing to develop a large LNG infrastructure, face economic difficulties in doing so. This raises the question of whether the states have the resources to develop mega-projects or energy infrastructure from scratch. This issue has technical, institutional and financial aspects.

Several lessons can be drawn from ten years of private participation in power generation:

■ the demand of a single state is too limited for large IPPs, as became evident with Dabhol. Several states should participate in such a project;

■ difficulties in finalising fuel supply agreements hampered some IPP projects based on coal;

■ poor co-ordination between the coal and power industries and the absence of a national grid able to transmit electric power from coal rich regions to markets made it difficult to launch large mine-mouth coal power.

Box 8   The Benefits of Regional Electric Co-operation and Integration

This excerpt from the handbook written by E7 member companies[1] (E7, 2000) describes the main benefits of integrating a power system at regional level. Though the book's focus is on a region comprising independent states, most of the issues addressed are also of relevance to India.

"For a given region, the integration of the electricity-supply industry of the member countries, in its final stage, may be defined by two objectives. First of all, the national electricity networks should be interconnected enough to enable substantial energy and capacity exchanges between countries. Then, having agreed on a certain level of quality of supply, the operators and the developers of the region's power systems have to co-ordinate in order to minimise the regional cost of electric power, while contributing to environmental and social objectives.

Generally speaking, regional electricity co-operation and integration enhances the contribution of the electricity sector to sustainable development. In OECD countries, further integrating large and mature electric power industries may yield important environmental and economic benefits. In developing and emerging economies, pooling electricity resources (notwithstanding political obstacles) is crucial for the development of the electricity-supply industry, as well as for the contribution of the industry to economic, environmental and social objectives, which are the three pillars of sustainable development.

Regional electricity co-operation and integration ranges from sharing experience and expertise on the operation and planning of the electric power system, to pooling activities such as training electric engineers, research and development, integrating parts of or the entire structure for operating and developing the electric power systems. The electric interconnection of national power systems is considered as a very important step toward regional electricity integration, and a decisive step toward the implementation of a regional competitive power market. There is, in every sector of the economy, particularly sectors of mass production, a clear case for pooling resources. This is all the more true in the electricity-supply industry: as electricity is not storable, there is a strong incentive for pooling supply and consumption through the interconnection of electricity networks.

Throughout the last century, the experience of utilities in the E7 countries has indicated that the interconnection of isolated electricity networks usually results in pooling generation resources and eventually, if the institutional structure permits, in integrating electric utilities into larger structures. Conversely, the existence of separate political and/or institutional structures may be an obstacle to technically and economically feasible electric interconnections, and may lead to the development of sub-optimal power systems at the expense of sustainable development objectives.

---

1. E7 is a gathering of electricity companies created in 1992 to share their knowledge and promote global environmental protection. In 2001, E7 members are: AEP, EDF, ENEL, KANSAI, OPG, RWE, SP, HQ, TEPCO. See www.e7.org for more information.

Last but not least, true regional integration will help to further optimise the use of generation resources. Through more efficient exploitation of hydroelectric resources and fossil fuels savings, it can also allow significant reductions of CO_2 and other airborne emissions. Accordingly, integration projects may benefit from the CDM, one of the flexible measures outlined in the Kyoto Protocol.

In developing countries, the important financial risk perceived by private investors combined with the scarce domestic financial resources force the electric power industry to call for the support from international funding institutions. These international funding institutions tend to favour regional co-operation projects *versus* separate national projects. They also urge the governments to reform their power sector toward more regulatory and financial independence, and to promote competition, wherever feasible. Interconnection projects will find the required funds more easily if they benefit financially independent and internationally accountable electric utilities. Conversely, an efficient electric power wholesale competition requires a minimum level of interconnection."

Integration of Political Decision-Making at the Federal Level

Private investment, especially foreign investment, favours large power projects. To attract investments to India, a sizeable power market must be developed. This requires, at the very least, pooling the power demand of neighbouring states to mitigate commercial risk. Technical and economic integration of state power markets would be facilitated by regional or federal integration of the institutional structure of the electric supply industry. This is all the more true because SEBs are allocated a budget and do not have the possibility for financial leveraging that central generating companies have. The states, which developed and still control more than 60% of Indian power generation capacity, are now political and institutional barriers to the development of a much-needed integrated power market.

Despite the willingness expressed over the past decade by successive Ministers of Power to advance fast-track projects and carry out a mega-power policy, many projects have been delayed or stalled by lack of agreement on fuel supply. Achieving fuel supply agreements would be facilitated if liquid fuels, gas and coal were controlled at the federal level by the same political entity as the one controlling the electric supply industry. In that case, clear and efficient co-ordination should be developed between the ministries concerned and the electric power industries.

The Indian electricity system as a whole should be planned, developed and operated as an integrated system. This would allow the development of an integrated long-term energy resource plan, as has been recommended by the World Bank for some time (World Bank, 1999). A minister for energy could be responsible for developing and supervising implementation of a comprehensive energy strategy.

Horizontal Integration of the Electric Supply Industry

The recent negotiation between states and the Government of India to find a solution for the payment of arrears owed by SEBs to central-sector power corporations demonstrates that central and state governments may not share the same understanding of key issues. State governments often are exploiting the fact that, under the Constitution, electricity is mainly their responsibility. This can lead to considerable delays in the implementation of reforms at the state level.

The basic administration unit in India remains the state, and each state runs its own utility. In the 1960s, a first step toward integration was taken by grouping the states into six regions. In 1975, the central government established generating companies, one for thermal power plants and another for hydroelectric power plants (NTPC and NHPC). Later, the national POWERGRID company was created. In recent years, NTPC developed its generating capacity faster than did the SEBs, which have been suffering from a deteriorating financial situation. This trend should be encouraged, along with competition between central generating companies.

The mega-project policy, which calls for the development of large generation projects to supply more than one state, is a good start toward integration of the Indian power sector. Of the 18 mega-projects announced however, few have shown much progress. For these to advance, more secure guarantees need to be provided, perhaps from the central government, and the whole procedure needs to be streamlined. All IPP projects, not just those announced as mega-projects, should be co-ordinated by a federal entity. And a single buyer should probably be responsible for power purchase agreements. PTC could be that entity.

THE POWER MIX

India is endowed with the third-largest coal reserves in the world; hence the bulk of electric power supply is naturally based on coal. Is the time ripe for the development of a large capacity based on natural gas combined-cycle units? Should India emphasise the development of nuclear energy or other very capital-intensive technology? Should India increase harnessing its huge hydroelectric resources in the North and East? What are the possibilities for renewables, especially wind energy?

Table 7

Projected Electricity Capacity and Generation in India

	1997		2020	
	GW	**TWh**	**GW**	**TWh**
Total	**103**	**464**	**308**	**1,484**
Coal	66	339	193	1,008
Oil	3	12	6	32
Gas	9	28	47	216
Nuclear	2	10	6	39
Hydro	22	75	50	171
Other Renewable	1	0	6	18

Source: IEA, 2000a.

The IEA projects a threefold rise in India's generation capacity from 1997 to 2020, which corresponds to an average yearly growth rate of 5.2%. The development of coal-fired generation based mostly on domestic resources will continue, with a threefold increase to 308 GW in 2020. There will is also be a fivefold increase in the use of natural gas to 47 GW.

In 1997, the load factor of the Indian power system was 51% – a low rate by international standards. This rate is assumed to increase to 55% in 2020. The most salient assumption is a sharp increase in the load factor of gas-fired power plants, from 36% to 53%, due mostly to the expected market deployment of natural gas combined-cycle units for baseload uses. The IEA also assumes a sizeable increase in the use of renewable energy.

Several conditions need to be met for these assumptions to become reality. The first is effective national integration of the state power systems. This would further facilitate the development of infrastructure for natural gas and the development of hydroelectric capacity. If integration does not occur, and without a major technology breakthrough, capacity additions would have to be limited to medium-size coal power plants for base and intermediate loads and oil-fired plants (using naphtha or diesel oil) for peak loads.

Coal

About 80% of the steam coal and virtually all the lignite produced in India are used for electric power generation, of which 73% is generated with coal. Coal is assumed to remain the fuel of choice for baseload power generation.

The reserve/production ratio is so favourable that there is no energy security concern. In 1998/1999, coal production was 292 million tonnes with estimated coal resources of 206 billion tonnes (bt).

Several issues need to be considered by the coal, railway and electric power industries:

- **the geographical concentration of coal.** Coal deposits are mostly situated in the eastern region (Bihar, 69 bt; Orissa, 50 bt; West Bengal, 26 bt), and in a lesser extent in the western region (Madhya Pradesh, 43 bt; and Maharashtra, 7 bt). Thirty coal trains per day travel from the East to the North and another 22 from the East and the Centre to the South. Whether to transport electricity over long distances, rather than coal, or importing coal, has to be studied on a long-term basis, taking into account long-term development costs, as well as environmental and social costs;

- **the poor quality of Indian coal.** More than 73% of the raw coal extracted has an ash content from 30% to 55%. In the Common Minimum Action Plan (1996), Chief Ministers decided to promote mega-power projects at pit heads and to set up coal washeries (to separate coal from dead-rock). Achieving these goals is proving difficult, especially due to the lack of co-ordination between the public authorities in charge of power and those responsible for coal or for transporting it by rail;

- **the lack of geological information on coal deposits.** The risk is high to invest in a mine-mouth power plant because of uncertainty about actual coal reserves;

■ **the possibility of using lignite instead of coal** because of constraints on rail transport to the South, increasing the use of southern lignite for power generation has been suggested. This would affect the electricity-supply industry, the mining industry and the railways.

To reduce the problems faced by IPPs, the Indian government should first formulate economic plans and a financial and organisational framework for co-ordinated development of the coal and electric-supply industries. The Planning Commission could supervise studies to determine the best infrastructure for sustainable development.

Investment in research and development by CSUs (NTPC in particular) should be strongly encouraged since it is still at low level. The Indian government should also co-operate with OECD countries in the development of environmentally-friendly coal technologies such as those that convert coal, oil refinery residues and other fuels into synthetic gas, and especially Integrated Gasification Combined-Cycle (IGCC) technology.

Natural Gas

Natural gas in combined-cycle power plants is often considered the fuel of choice for baseload generation. The advantages of this technology (series effect, wide range of unit sizes, low environmental impact, short lead times, competitive investment cost, ease of operation) could accelerate shifting from the traditional model of a large vertically-integrated public monopoly to a competitive power market with decentralised generation units.

In India, however, power plant manufacturers are still more familiar with coal-boiler and steam-turbine technology than with combustion-turbine technology. Moreover, India does not yet have access to cheap natural gas resources.

For these reasons, the Indian government should promote only a selected number of LNG-based power-generation projects and clarify the environment of development for the numerous projects of LNG terminals. Likewise, there will be no real incentive for large-scale development of natural gas combined cycle technology unless:

■ breakthroughs in LNG technology dramatically reduce the cost, and/or;

■ delays and problems in the co-ordination of the electric-supply and coal industries continue to hamper project development.

The situation could change if local and global environmental concerns were further integrated, leading to the adoption and implementation by India of more stringent emission control measures. Bangladesh has sizeable domestic natural gas resources and a short- to medium-term surplus in generation capacity, further integration of natural gas markets in South Asia could also change the picture.

India as a subcontinent with one-sixth of the world's population should keep open all technology options for power generation. In coastal regions, especially those closest to the Middle East and Central Asia, natural gas may become competitive with domestic coal for baseload power generation.

Map 4 Coal Production, Use and Imports

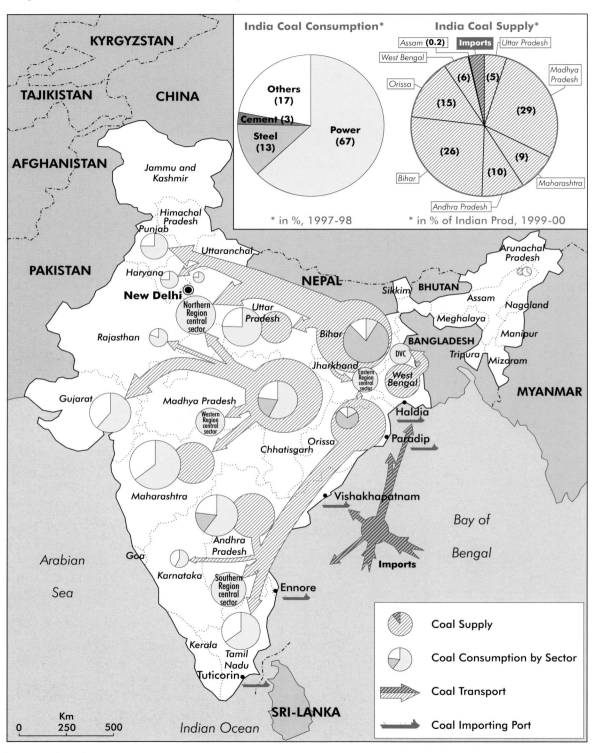

Sources: IEA analysis.

Map 5 Main Projects of LNG Terminals

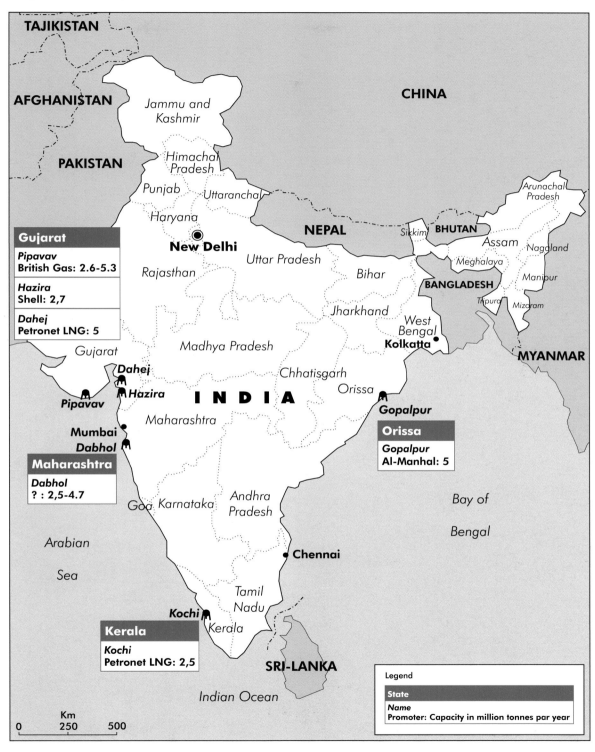

Gujarat

Pipavav
British Gas: 2.6-5.3

Hazira
Shell: 2,7

Dahej
Petronet LNG: 5

Orissa

Gopalpur
Al-Manhal: 5

Maharashtra

Dabhol
? : 2,5-4.7

Kerala

Kochi
Petronet LNG: 2,5

Legend

State

Name
Promoter: Capacity in million tonnes par year

Source: IEA, 2000a; TERI, 2000.

Nuclear Energy   The situation for nuclear energy is about the same as for natural gas: nuclear is at an economic disadvantage compared to coal for baseload generation. Furthermore, the investment required to develop a competitive nuclear-power industry would be even greater than for LNG technology.

For the coming years, nuclear technology should remain the responsibility of the central government, with strong emphasis on international co-operation in the development of innovative nuclear reactors and fuel cycles.

Hydroelectric Energy   In 1997, India generated 75 TWh using hydroelectric energy, compared with a total hydroelectric potential of 600 TWh.

Table 8   Regional Hydroelectric Potential and Energy Requirements

	Hydroelectricity potential	Hydroelectricity to be developed		Total electricity requirements, 2011-2012
	TWh	TWh	% of total	TWh
Northern	225	193	37%	350
Western	31	21	4%	321
Southern	62	31	6%	234
Eastern	43	36	7%	135
North-Eastern	239	237	46%	18
Total	**600**	**518**	**100%**	**1,058**

Source: CEA, 1997.

Table 8 shows that the hydroelectric potential is not evenly located, with 83% in the North and North-East, in the Brahmaputra, Indus and Ganges river basins. State-by-state analysis would show even greater disparity between load demand and hydroelectric potential.

In the last ten years, the development of hydroelectric power, which was expected to represent 40% of generation capacity, actually slowed down. Hydropower developed more slowly than did thermal power. Hydro capacity has barely doubled since 1980, while thermal capacity has nearly quadrupled. Hydro's share in the power-plant mix has declined, with negative implications for peak-load availability.

Various aspects of hydroelectricity explain this situation:

■ hydroelectricity is the most capital-intensive of all power generation technologies;

■ the most economically viable sites are often the biggest;

■ the time lag between feasibility studies and commissioning is very long;

■ the development of hydroelectric projects is further complicated by environmental requirements and public opinion (resettling displaced populations in densely populated areas is not an easy task);

- geological risks may be considerable;

- operational flexibility cannot be guaranteed for peak load;

- hydrological risk is hard to manage, water flows may vary dramatically from year to year;

- due to the size and concentration of India's hydroelectric potential, bulk interstate power transmission lines would need to be built from remote hydroelectric projects to load centres;

- it is hard to develop hydroelectric potential in a liberalised power market because private investors are reluctant to back projects that are prone to geological contingencies and where most of the costs have to be paid upfront.

Due to these problems, the government will have to play a dynamic and driving role in harnessing hydroelectric potential. Its efforts should focus on:

- developing the central transmission utility's investments in interstate transmission system;

- making hydroelectric projects attractive for private investors;

- improving the performance of central generating companies developing hydroelectricity.

To encourage private investors, hydroelectric plants could be developed step by step, as their unit investment cost is commensurate with that of combined-cycle thermal plants.

Renewables

The total commercially exploitable potential from renewables is estimated at about 47,000 MW: 20,000 MW from wind, 10,000 MW from small hydro, and 17,000 MW from biomass/bioenergy. The government is promoting renewables with increasing allocations in its five-year plans.

But renewables still have only a negligible share of total commercial primary energy in India (2.5%, including hydro in 1998). Nonetheless, their share is growing and translates into large absolute numbers, given the size of the Indian energy sector. As a result, India is emerging as a world leader in the diffusion and development of several renewable energy technologies.

Installed wind-power capacity, which totalled about 1,200 MW in 2000, is among the highest in the world. It increased rapidly in the 1990s, boosted by subsidies and financial incentives. Its projected rise to 4 GW by 2020 (IEA) will require an even stronger government commitment. One initiative is a proposal to introduce a fossil-fuel levy to fund the development of renewables.

India's solar potential is also large and is being tapped for heating and photovoltaïc power. A 140 MW Integrated Solar Combined-Cycle power plant is under construction in Rajasthan.

Table 9   Renewable Energy Development in India, as of 30 June 2000

Source / Technologies	Units	Amount	World Rank
Power Generation			
Wind power	MW	1,175	5
Small hydro power (<25MW)	MW	1,157	10
Biomass-based power	MW	235	4
Biomass gasifiers	MW	35	1
Solar photovoltaics	MW	58	3
Energy recovery from urban & industrial wastes	MWe	15	na
Thermal Applications			
Biogas plants	No.	3,043,853	2
Improved cookstoves	No.	32,267,000	2
Solar water heating systems	m2	500,000	
Solar cookers	No.	490,000	1
Water pumping			
Wind pumps	No.	651	na
Solar PV pumps	No.	3,443	na

Source: Ministry of Non-Conventional Energy Sources

Policy goals to accelerate market deployment of renewable energy are being formulated in India. The *Policy Statement on Renewable Energy* that the Ministry of Non-conventional Energy Sources began drafting in 2000 is India's first attempt to develop a comprehensive renewable energy policy. It will define policy goals, as well as identify mechanisms and investments required to achieve them. The objectives are to meet minimum rural energy needs; provide decentralised/off-grid energy supply for the agriculture, industry, commercial and household sectors in rural and urban areas, and generate and supply grid-quality power. The medium-term goals, to 2012, call for:

■ 10% of new power capacity to come from renewables;

■ progressive electrification by renewables of the 18,000 villages considered non-electrifiable by conventional means;

■ improved woodstoves in 30 million households;

■ three million additional family-size biogas plants;

■ five million solar lanterns and two million solar home-lighting systems;

■ solar water heating systems in one million homes.

These goals could be achieved by taking a market approach to renewable energy development and moving away from a purely product-based approach to one that delivers specific services for different markets such as: grid power, decentralised power (distributed generation), rural energy (often off-grid). Policy measures will differ for each market segment. Connecting renewable power to the grid requires liberalisation

Map 6 Wind Resources in Ten states of India, at 50 Meters Above Ground

TAJIKISTAN

AFGHANISTAN

PAKISTAN

CHINA

Jammu and Kashmir

Himachal Pradesh

Punjab

Uttaranchal

Haryana

⊙
New Delhi

Rajasthan

NEPAL

Sikkim BHUTAN

Arunachal Pradesh

Assam Nagaland

Meghalaya Manipur

Uttar Pradesh

Bihar

BANGLADESH

Tripura Mizoram

MYANMAR

Jharkhand

West Bengal

Madhya Pradesh

Gujarat

Chhatisgarh

Orissa

I N D I A

Maharashtra

Bay of

Bengal

Andhra Pradesh

Goa

Karnataka

Arabian

Sea

Tamil Nadu

Kerala

SRI-LANKA

Km
0   250   500

Indian Ocean

■	350-450 W/m²
▨	300-350 W/m²
▨	250-300 W/m²
□	200-250 W/m²

Source: TERI, 2000.

of the power industry, particularly tariffs. The draft Electricity Bill, which is expected to provide a legal framework for integrating power development in India, should include provisions for market deployment of renewable energy. So far, the policy implications of renewable energy are not clearly identified in the Electricity Bill. Indian regulators at both central and state levels have yet to take full account of renewable power's special requirements in their tariff orders or in the norms they have begun setting.

REFERENCES

Ahluwalia, S., 2000, "Power Tariff Reform in India", *Economic and Political Weekly*, September 16, 2000.

Alam M., Sathaye J., Barnes D., 1998, "Urban household energy use in India: efficiency and policy implications", *Energy Policy*, 26/11, pp. 885-891.

Angus D., 2001, "Power failure hits North India", *Financial Times*, January 3, 2001.

CEA, 1997, *Fourth National Power Plan 1997-2012*, Central Electricity Authority, New Delhi.

CEA, 1998, *Fuel map of India*, Central Electricity Authority, New Delhi.

CEA, 1998a, *Average Electric Rates and Duties in India*, Central Electricity Authority, New Delhi.

CEA, 1999, *Perspective Transmission Plan 2011-2012*, Central Electricity Authority, New Delhi.

CMIE, 2001, *Energy*, Centre for Monitoring Indian Economy, Mumbai.

D'Sa A., Narasimha Murthy K.V., Reddy A.K.N., 1999, "India's Power Sector Liberalisation", *Economic and Political Weekly*, June 5, 1999.

DERC, 2000, *Concept paper on tariff*, Delhi Electricity Regulatory Commission, New Delhi.

ESMAP, 2000, *Energy Services for the World's Poor, Energy and Development Report 2000*, World Bank, Washington.

ESMAP & Energy and Mining Sector Board, 2001, *California Power Crisis: Lessons for Developing Countries*, World Bank, Washington.

Godbole M. & al., 2001, *Report of the Energy Review Committee*, Part I and Part II, Government of Maharashtra, Mumbai.

GOI, 1996, *Common Minimum National Action Plan for Power*, Ministry of Power, New Delhi.

GOI, 1999, *Annual report on the working of State Electricity Boards & Electric Departments*, Planning Commission, New Delhi.

GOI, 2000, *Annual report on the working of State Electricity Boards & Electric Departments*, Planning Commission, New Delhi.

GOI, 2000a, *Economic Survey 1999-2000*, Ministry of Finances, New Delhi.

GOI, 2001, *Blueprint for power sector development*, Ministry of Power, New Delhi.

GOI, 2001a, *Annual report on the working of State Electricity Boards & Electric Departments*, Planning Commission, New Delhi.

GOI, 2001b, *Annual Report 2000-2001*, Ministry of Power, New Delhi.

Hautot A., 1999, "A method for overall costs comparison: analysis of the main cost drivers", *Workshop on Electricity Network Tariffs*, Prague 19-20 May 1999.

IEA, 1998, *World Energy Outlook 1998*, International Energy Agency, Paris.

IEA, 1998a, *Natural gas pricing in competitive markets*, International Energy Agency, Paris.

IEA, 1999, *World Energy Outlook, Insights 1999, Looking at Energy Subsidies*, International Energy Agency, Paris.

IEA, 1999b, *Electricity Market Reform, an IEA handbook*, International Energy Agency, Paris.

IEA, 1999c, *Electricity Reform, Power generation costs and investment*, International Energy Agency, Paris.

IEA, 2000a, *World Energy Outlook 2000*, International Energy Agency, Paris.

IEA, 2001, *Energy Balances of Non-OECD Countries, 1998-1999*, International Energy Agency, Paris.

IEA, 2001a, *Key world energy statistics*, International Energy Agency, Paris.

IEA, 2002, *Russia Energy Survey 2002*, International Energy Agency, Paris.

Maggo J.N., 1998, *Sectoral energy demand in the Ninth-Plan and the perspective period up to 2011-12, A technical note*, Planning Commission, New Delhi.

Morris S., 1996, "Political Economy of Electric Power in India", *Economic and Political Weekly*, May 18, 1996.

PGCIL, 1999, *Indian electricity grid code*, Power Grid Corporation of India Ltd, New Delhi.

Rao S.L., 2000, "Power Tariff blues of Andhra Pradesh", *Journal of the Council of Power Utilities*, Vol. VIII, 3, July-September 2000.

Rao S.L., 2001, "Dabhol, Godbole Report and the Future", *Economic and Political Weekly*, May 12, 2001.

RSEB, 1999, *Revised Rates*, Rajasthan State Electricity Board, Jaipur.

Ruet J., 2001, "Winners and Losers of the SEB Reform: An Organisational Overview", *CSH Occasional Paper* 1, Centre de Sciences Humaines, New Delhi.

Sankar T.L., Ramachandra U., 2000, "Electricity Tariffs Regulator, The Orissa Experience", *Economic and Political Weekly*, May 27, 2000.

Srivastava R.N, Sinha K.N., Goel R.S., 1998, "Planning power development in India – emphasis on hydro projects", *17th World Energy Congress*, Houston, September 1998.

TERI, 2000, *Energy Data Directory & Year Book 2000-2001*, Tata Energy Research Institute, New Delhi.

WEC, 2000, *Renewable Energy in South Asia, Status and Prospects*, World Energy Council / South Asian Association for Regional Co-operation, Colombo.

WEC, 2001, *Electricity Market Design and Creation in Asia Pacific*, World Energy Council, London.

World Bank, 1996, *Staff Appraisal report India Orissa power sector restructuring project*, World Bank, Washington.

World Bank, 1999, "Fuelling India's growth and development", *South Asia Brief*, July.

World Bank, 1999a, *Project Appraisal document on a proposed loan in the amount of US$210 million equivalent to India for Andhra Pradesh power sector restructuring program*, World Bank, Washington.

World Bank, 1999b, *Meeting India's energy needs (1978-1999), A Country Sector Review*, Operation Evaluation Department, World Bank, Washington.

World Bank, 1999c, *Meeting India's Future Power Needs – Planning for Sustainable Development. Environmental Issues in the Power Sector*, World Bank, Washington.

World Bank, 2000, *Project Appraisal document on a proposed loan in the amount of US$ 150 million to the Government of India for the Uttar Pradesh power sector restructuring project*, World Bank, Washington.

World Bank, 2000a, *India – Country Framework Report for Private Sector Participation in Infrastructure*, World Bank, Washington.

Wu Z., Soud H., 1998, *Air pollution control and coal-fired power generation in the Indian subcontinent*, IEA Coal Research Centre, London.

ANNEX 1 IEA SHARED GOALS

The 26 Member countries of the International Energy Agency (IEA) seek to create the conditions in which the energy sectors of their economies can make the fullest possible contribution to sustainable economic development and the well-being of their people and of the environment. In formulating energy policies, the establishment of free and open markets is a fundamental point of departure, though energy security and environmental protection need to be given particular emphasis by governments. IEA countries recognise the significance of increasing global interdependence in energy. They therefore seek to promote the effective operation of international energy markets and encourage dialogue with all participants. In order to secure their objectives they therefore aim to create a policy framework consistent with the following goals:

1. **Diversity, efficiency and flexibility within the energy sector** are basic conditions for longer-term energy security: the fuels used within and across sectors and the sources of those fuels should be as diverse as practicable. Non-fossil fuels, particularly nuclear and hydro power, make a substantial contribution to the energy supply diversity of IEA countries as a group.

2. Energy systems should have **the ability to respond promptly and flexibly to energy emergencies**. In some cases this requires collective mechanisms and action: IEA countries co-operate through the Agency in responding jointly to oil supply emergencies.

3. **The environmentally sustainable provision and use of energy** is central to the achievement of these shared goals. Decision-makers should seek to minimise the adverse environmental impacts of energy activities, just as environmental decisions should take account of the energy consequences. Government interventions should where practicable have regard to the Polluter Pays Principle.

4. **More environmentally acceptable energy sources** need to be encouraged and developed. Clean and efficient use of fossil fuels is essential. The development of economic non-fossil sources is also a priority. A number of IEA members wish to retain and improve the nuclear option for the future, at the highest available safety standards, because nuclear energy does not emit carbon dioxide. Renewable sources will also have an increasingly significant contribution to make.

5. **Improved energy efficiency** can promote both environmental protection and energy security in a cost-effective manner. There are significant opportunities for greater energy efficiency at all stages of the energy cycle from production to consumption. Strong efforts by Governments and all energy users are needed to realise these opportunities.

6. **Continued research, development and market deployment of new and improved energy technologies** make a critical contribution to achieving the objectives outlined above. Energy technology policies should complement broader energy policies. International co-operation in the development and dissemination of energy technologies, including industry participation and co-operation with non-Member countries, should be encouraged.

7. **Undistorted energy prices** enable markets to work efficiently. Energy prices should not be held artificially below the costs of supply to promote social or industrial goals. To the extent necessary and practicable, the environmental costs of energy production and use should be reflected in prices.

8. **Free and open trade** and a secure framework for investment contribute to efficient energy markets and energy security. Distortions to energy trade and investment should be avoided.

9. **Co-operation among all energy market participants** helps to improve information and understanding, and encourage the development of efficient, environmentally acceptable and flexible energy systems and markets worldwide. These are needed to help promote the investment, trade and confidence necessary to achieve global energy security and environmental objectives.

IEA Ministers adopted the "Shared Goals" at their 4 June 1993 meeting in Paris.

ANNEX 2

INDIA, ENERGY BALANCES AND KEY STATISTICAL DATA

Unit: Mtoe

SUPPLY	1973	1990	1992	1994	1996	1998	1999
TOTAL PRODUCTION	**177.30**	**333.82**	**351.04**	**368.82**	**394.21**	**409.72**	**409.79**
Coal[1]	39.86	106.06	119.62	127.97	144.33	150.41	147.28
Oil	7.36	34.04	28.49	31.58	33.71	33.65	33.24
Gas	0.63	10.13	13.08	13.92	17.30	20.01	20.75
Comb. Renewables & Wastes[2]	126.34	175.82	182.08	186.74	190.47	195.28	198.02
Nuclear	0.62	1.60	1.75	1.47	2.36	3.09	3.41
Hydro	2.49	6.16	6.01	7.11	5.93	7.18	7.00
Geothermal	–	–	–	–	–	–	–
Solar/Wind/Other[3]	–	0.00	0.00	0.02	0.10	0.09	0.09
TOTAL NET IMPORTS[4]	**16.81**	**29.09**	**40.93**	**46.07**	**60.07**	**66.25**	**69.92**
Coal[1] Exports	0.25	0.05	0.06	0.05	0.06	0.16	0.38
Imports	0.00	3.14	4.15	7.00	8.80	9.62	11.13
Net Imports	–0.24	3.09	4.08	6.95	8.74	9.46	10.76
Oil Exports	0.19	2.37	4.17	3.47	3.90	3.42	2.60
Imports	17.47	28.67	41.36	43.00	55.22	60.21	61.76
Bunkers	0.23	0.41	0.45	0.53	0.12	0.10	0.09
Net Imports	17.06	25.89	36.74	38.99	51.20	56.70	59.07
Gas Exports	–	–	–	–	–	–	–
Imports	–	–	–	–	–	–	–
Net Imports	–	–	–	–	–	–	–
Electricity Exports	0.00	0.01	0.01	0.00	0.01	0.03	0.03
Imports	–	0.12	0.12	0.13	0.14	0.12	0.12
Net Imports	–0.00	0.12	0.10	0.12	0.13	0.09	0.09
TOTAL STOCK CHANGES	**–0.39**	**–3.80**	**–5.00**	**–0.29**	**3.24**	**–4.63**	**0.71**
TOTAL SUPPLY (TPES)	**193.72**	**359.11**	**386.97**	**414.59**	**457.52**	**471.34**	**480.42**
Coal[1]	39.41	106.55	122.41	136.97	160.50	160.29	157.17
Oil	24.24	58.74	61.53	68.23	80.73	85.30	93.88
Gas	0.63	10.13	13.08	13.92	17.30	20.01	20.75
Comb. Renewables & Wastes[2]	126.34	175.82	182.08	186.74	190.47	195.28	198.02
Nuclear	0.62	1.60	1.75	1.47	2.36	3.09	3.41
Hydro	2.49	6.16	6.01	7.11	5.93	7.18	7.00
Geothermal	–	–	–	–	–	–	–
Solar/Wind/Other[3]	–	0.00	0.00	0.02	0.10	0.09	0.09
Electricity Trade[5]	–0.00	0.12	0.10	0.12	0.13	0.09	0.09
Shares (%)							
Coal	*20.3*	*29.7*	*31.6*	*33.0*	*35.1*	*34.0*	*32.7*
Oil	*12.5*	*16.4*	*15.9*	*16.5*	*17.6*	*18.1*	*19.5*
Gas	*0.3*	*2.8*	*3.4*	*3.4*	*3.8*	*4.2*	*4.3*
Comb. Renewables & Wastes	*65.2*	*49.0*	*47.1*	*45.0*	*41.6*	*41.4*	*41.2*
Nuclear	*0.3*	*0.4*	*0.5*	*0.4*	*0.5*	*0.7*	*0.7*
Hydro	*1.3*	*1.7*	*1.6*	*1.7*	*1.3*	*1.5*	*1.5*
Geothermal	*–*	*–*	*–*	*–*	*–*	*–*	*–*
Solar/Wind/Other	*–*	*–*	*–*	*–*	*–*	*–*	*–*
Electricity Trade	*–*	*–*	*–*	*–*	*–*	*–*	*–*

O is negligible, – is nil, ... is not available.

Unit: Mtoe

DEMAND

FINAL CONSUMPTION BY SECTOR

	1973	1990	1992	1994	1996	1998	1999
TFC	**40.43**	**113.09**	**125.98**	**327.47**	**354.10**	**357.42**	**360.90**
Coal[1]	14.61	37.61	40.31	44.97	52.23	38.83	30.37
Oil	20.76	51.98	58.33	64.25	76.16	83.54	90.14
Gas	0.33	4.97	5.67	6.27	7.23	9.14	9.84
Comb. Renewables & Wastes[2]	–	–	–	186.74	190.47	195.28	198.02
Geothermal	–	–	–	–	–	–	–
Solar/Wind/Other	–	–	–	–	–	–	–
Electricity	4.73	18.53	21.67	25.24	28.01	30.62	32.53
Heat	–	–	–	–	–	–	–
Shares (%)							
Coal	36.1	33.3	32.0	13.7	14.8	10.9	8.4
Oil	51.4	46.0	46.3	19.6	21.5	23.4	25.0
Gas	0.8	4.4	4.5	1.9	2.0	2.6	2.7
Comb. Renewables & Wastes	–	–	–	57.0	53.8	54.6	54.9
Geothermal	–	–	–	–	–	–	–
Solar/Wind/Other	–	–	–	–	–	–	–
Electricity	11.7	16.4	17.2	7.7	7.9	8.6	9.0
Heat	–	–	–	–	–	–	–
TOTAL INDUSTRY[6]	**20.93**	**64.57**	**70.09**	**100.43**	**114.53**	**106.79**	**102.47**
Coal[1]	9.29	34.40	37.63	44.35	51.88	38.67	30.11
Oil	8.14	15.93	16.60	17.33	21.23	24.09	26.78
Gas	0.31	4.84	5.40	5.98	6.86	8.73	9.17
Comb. Renewables & Wastes[2]	–	–	–	21.25	21.70	22.25	22.56
Geothermal	–	–	–	–	–	–	–
Solar/Wind/Other	–	–	–	–	–	–	–
Electricity	3.20	9.40	10.45	11.52	12.87	13.05	13.86
Heat	–	–	–	–	–	–	–
Shares (%)							
Coal	44.4	53.3	53.7	44.2	45.3	36.2	29.4
Oil	38.9	24.7	23.7	17.3	18.5	22.6	26.1
Gas	1.5	7.5	7.7	5.9	6.0	8.2	8.9
Comb. Renewables & Wastes	–	–	–	21.2	18.9	20.8	22.0
Geothermal	–	–	–	–	–	–	–
Solar/Wind/Other	–	–	–	–	–	–	–
Electricity	15.3	14.6	14.9	11.5	11.2	12.2	13.5
Heat	–	–	–	–	–	–	–
TRANSPORT[7]	**11.91**	**26.44**	**30.88**	**33.17**	**39.19**	**41.68**	**44.47**
TOTAL OTHER SECTORS[8]	**7.59**	**22.08**	**25.02**	**193.88**	**200.38**	**208.95**	**213.96**
Coal[1]	1.17	0.72	0.61	0.30	0.29	0.15	0.25
Oil	5.01	12.46	13.35	14.58	16.37	18.42	19.58
Gas	0.02	0.13	0.27	0.30	0.37	0.41	0.68
Comb. Renewables & Wastes[2]	–	–	–	165.49	168.77	173.04	175.46
Geothermal	–	–	–	–	–	–	–
Solar/Wind/Other	–	–	–	–	–	–	–
Electricity	1.40	8.78	10.79	13.21	14.57	16.94	18.00
Heat	–	–	–	–	–	–	–
Shares (%)							
Coal	15.4	3.2	2.4	0.2	0.1	0.1	0.1
Oil	66.0	56.4	53.4	7.5	8.2	8.8	9.1
Gas	0.3	0.6	1.1	0.2	0.2	0.2	0.3
Comb. Renewables & Wastes	–	–	–	85.4	84.2	82.8	82.0
Geothermal	–	–	–	–	–	–	–
Solar/Wind/Other	–	–	–	–	–	–	–
Electricity	18.4	39.7	43.1	6.8	7.3	8.1	8.4
Heat	–	–	–	–	–	–	–

Unit: Mtoe

DEMAND

ENERGY TRANSFORMATION AND LOSSES

	1973	1990	1992	1994	1996	1998	1999
ELECTRICITY GENERATION[9]							
INPUT (Mtoe)	26.04	70.72	83.22	93.13	110.40	127.08	132.93
OUTPUT (Mtoe)	**6.26**	**24.89**	**28.61**	**33.16**	**37.56**	**42.69**	**45.35**
(TWh gross)	72.80	289.44	332.71	385.53	436.70	496.39	527.33
Output Shares (%)							
Coal	50.3	67.6	70.9	71.0	75.3	76.3	75.2
Oil	6.1	2.3	1.9	1.5	1.5	1.1	1.1
Gas	0.5	3.2	4.1	4.5	5.1	3.2	5.5
Comb. Renewables & Wastes	–	–	–	–	–	–	–
Nuclear	3.3	2.1	2.0	1.5	2.1	2.4	2.5
Hydro	39.8	24.8	21.0	21.5	15.8	16.8	15.4
Geothermal	–	–	–	–	–	–	–
Solar/Wind/Other	–	0.0	0.0	0.0	0.3	0.2	0.2
TOTAL LOSSES	**27.14**	**67.60**	**78.64**	**85.58**	**99.71**	**113.86**	**120.06**
of which:							
Electricity and Heat generation[10]	19.78	45.82	54.61	59.97	72.84	84.39	87.58
Other Transformation	3.97	10.21	11.64	11.96	11.93	11.46	13.04
Own Use and Losses[11]	3.39	11.56	12.39	13.64	14.94	18.01	19.44
Statistical Differences	**126.15**	**178.43**	**182.35**	**1.54**	**3.72**	**0.05**	**–0.55**
INDICATORS	**1973**	**1990**	**1992**	**1994**	**1996**	**1998**	**1999**
GDP (billion 1995 US$)	120.92	275.10	291.24	328.11	377.78	421.71	449.12
Population (millions)	586.22	849.52	882.30	913.60	945.61	979.67	997.52
TPES/GDP[12]	1.60	1.31	1.33	1.26	1.21	1.12	1.07
Energy Production/TPES	0.92	0.93	0.91	0.89	0.86	0.87	0.85
Per Capita TPES[13]	0.33	0.42	0.44	0.45	0.48	0.48	0.48
Oil Supply/GDP[12]	0.20	0.21	0.21	0.21	0.21	0.20	0.21
TFC/GDP[12]	0.33	0.41	0.43	1.00	0.94	0.85	0.80
Per Capita TFC[13]	0.07	0.13	0.14	0.36	0.37	0.36	0.36
Energy-related CO_2 emissions (Mt CO_2)[14]	217.4	591.1	667.6	745.9	880.0	893.5	903.8
CO_2 emissions from bunkers (Mt CO_2)	3.5	6.7	6.4	7.7	7.3	6.9	7.0
GROWTH RATES (% per year)	**73-79**	**79-90**	**90-92**	**92-94**	**94-96**	**96-98**	**98-99**
TPES	3.4	3.8	3.8	3.5	5.1	1.5	1.9
Coal	5.4	6.4	7.2	5.8	8.2	–0.1	–1.9
Oil	4.9	5.6	2.3	5.3	8.8	2.8	10.1
Gas	14.1	19.8	13.7	3.2	11.5	7.5	3.7
Comb. Renewables & Wastes	2.3	1.8	1.8	1.3	1.0	1.3	1.4
Nuclear	3.1	7.1	4.7	–8.4	26.7	14.4	10.2
Hydro	7.8	4.2	–1.2	8.8	–8.7	10.0	–2.4
Geothermal	–	–	–	–	–	–	–
Solar/Wind/Other	–	–	27.5	87.1	149.4	–4.3	–
TFC	5.4	6.7	5.5	61.2	4.0	0.5	1.0
Electricity Consumption	7.1	9.0	8.2	7.9	5.3	4.6	6.2
Energy Production	3.4	4.0	2.5	2.5	3.4	1.9	0.0
Net Oil Imports	3.0	2.2	19.1	3.0	14.6	5.2	4.2
GDP	3.2	5.9	2.9	6.1	7.3	5.7	6.5
Growth in the TPES/GDP Ratio	0.2	–2.0	0.9	–2.5	–2.1	–3.9	–4.3
Growth in the TFC/GDP Ratio	2.2	0.7	2.6	51.9	–3.1	–4.9	–5.2

Please note: Rounding may cause totals to differ from the sum of the elements.

Footnotes to Energy Balances and Key Statistical Data

1. Includes lignite and peat, except for Finland, Ireland and Sweden. In these three cases, peat is shown separately.
2. Comprises solid biomass, liquid biomass, biogas, industrial waste and municipal waste. Data are often based on partial surveys and may not be comparable between countries.
3. Other includes tide, wave and ambient heat used in heat pumps.
4. Total net imports include combustible renewables and waste.
5. Total supply of electricity represents net trade. A negative number indicates that exports are greater than imports.
6. Includes non-energy use.
7. Includes less than 1% non-oil fuels.
8. Includes residential, commercial, public service and agricultural sectors.
9. Inputs to electricity generation include inputs to electricity, CHP and heat plants. Output refers only to electricity generation.
10. Losses arising in the production of electricity and heat at public utilities and autoproducers. For non-fossil-fuel electricity generation, theoretical losses are shown based on plant efficiencies of 33% for nuclear, 10% for geothermal and 100% for hydro.
11. Data on "losses" for forecast years often include large statistical differences covering differences between expected supply and demand and mostly do not reflect real expectations on transformation gains and losses.
12. Toe per thousand US dollars at 1995 prices and exchange rates.
13. Toe per person.
14. "Energy-related CO_2 emissions" specifically means CO_2 from the combustion of the fossil fuel components of TPES (i.e. coal and coal products, peat, crude oil and derived products and natural gas), while CO_2 emissions from the remaining components of TPES (i.e. electricity from hydro, other renewables and nuclear) are zero. Emissions from the combustion of biomass-derived fuels are not included, in accordance with the IPCC greenhouse gas inventory methodology. Also in accordance with the IPCC methodology, emissions from international marine and aviation bunkers are not included in national totals. Projected emissions for oil and gas are derived by calculating the ratio of emissions to energy use for 1999 and applying this factor to forecast energy supply. Future coal emissions are based on product-specific supply projections and are calculated using the IPCC/OECD emission factors and methodology.

ANNEX 3

CENTRAL PUBLIC SECTOR GENERATING COMPANIES

■ NTPC, National Thermal Power Corporation Ltd

Incorporated in 1975 as a wholly government-owned enterprise with the objective of planning, promoting and organising integrated development of thermal power. NTPC generates power in all four major power regions of the country and contributed 26% of total power generation in the country during 2000-2001. Its approved capacity of 22,955 MW consists of thirteen coal stations and seven gas/liquid-fuel combined-cycle power plants. NTPC also manages the Government of India's Badarpur thermal power station (705 MW) of and the Balco Captive Power Plant (270 MW). In 2000-2001, NTPC's turnover was 192,200 million rupees.

http://www.ntpc.co.in

■ NHPC, National Hydro Power Corporation Ltd

Incorporated in November 1975 as a central government enterprise to undertake all activities from design to commissioning of hydro projects. NHPC included wind and tidal power among its projects in 1998 and geo-thermal and gas power in 1999 and is also preparing to take up mini/micro hydro projects. NHPC presently has an installed capacity of 2,175 MW from eight hydropower stations, and is engaged in the construction of six projects amounting to a total installed capacity of 2,280 MW. NHPC has drawn up a massive plan to add over 49,000 MW of hydropower capacity in the next 20 years. In 2000-2001, NHPC's turnover was 12 million rupees and its power stations generated 9,581 million kWh.

http://www.nhpcindia.com

■ NEEPCO, North Eastern Electric Power Corporation Ltd

Incorporated on 2 April 1976 as a wholly-owned Government of India enterprise to generate, transmit, operate, maintain and develop power stations in the entire North Eastern Region. NEEPCO currently manages three power projects (commissioned capacity: 625 MW – hydro 250 MW and gas 291 + 84 MW) and plans to add 3,515 MW of capacity in the next decade. In 1999, it generated 2,415 million units.

http://www.neepco.com

■ NLC, Neyveli Lignite Corporation Ltd

Registered as a company in November 1956, NLC Ltd exploits lignite deposits and generates lignite-based power. Its main units are lignite mines, thermal power stations, fertilizer plant and briquetting & carbonisation plants. Mine-I (6.5 million tonnes of lignite per annum) feeds Thermal Power Station-I (600 MW), Briquetting & Carbonisation Plant (262,000 tonnes of coke-achievable capacity) and the Process Steam Plant. Mine-II (10.5 MT of lignite per annum) feeds its captive Thermal Power Station-II (7 × 210 MW). The power generated from TPS-I is fed into the TNEB grid, which is the sole beneficiary. Power generated from TPS-II is shared by southern states (Tamil Nadu, Kerala, Karnataka, Andhra Pradesh and the Union Territory of Pondicherry). NLC is under the administrative control of the Ministry of Coal. In 1999-2000, NLC's turnover was 14,962 million rupees.

http://www.nlcindia.co.in

■ NPCIL, Nuclear Power Corporation of India Ltd

Registered in September 1987 as a wholly owned enterprise of the Government of India under administrative control of the Department of Atomic Energy to design, construct, operate and maintain atomic power stations for the Government of India. The company operates six nuclear power stations (generating 16,621 TWh in 2000-2001) and is constructing two nuclear power plants and handling other related activities consistent with the policies of the Government of India. In 1998-1999, NPCIL's turnover was 21,177 million rupees.

http://www.npcil.org

ANNEX 4

STATUS OF ELECTRICITY REFORMS IN THE INDIAN STATES, AS OF JUNE 2001

	State's Decision to Reform Power Sector	MoU with GOI	Reform Bill		SERC			SEBs Unbundling		Distribution Privatization		Financial Assistance
			Proposed	Enacted	Decision taken	Constituted	Functional	Envisaged	Done	Planned	Done	
Andra Pradesh	x	x		x			x		x	x		WB, DFID, CIDA
Assam	x	x			x							PFC
Bihar	x											
Delhi	x			x		x				x		
Gujarat	x	x	x				x	x				ADB
Goa	x				x							PFC
Haryana	x	x		x			x		x			WB, DFID
Himachal Pradesh	x	x					x					
Jammu & Kashmir	x											
Karnataka	x	x		x			x		x	x		WB
Kerala	x					x		Profit Centre Approach				CIDA
Madhya Pradesh	x	x		x			x	x				ADB
Maharashtra	x	x	x				x	x				PFC
Orissa	x			x			x		x		x	WB
Punjab	x	x				x						PFC, ADB
Rajasthan	x	x		x			x		x	x		WB
Tamil Nadu	x					x						PFC
Uttar Pradesh	x	x		x			x		X			WB
West Bengal	x	x					x					PFC

Sources: GOI, 2001b.
Note: this table does not include information about the Northeastern states of Arunachal Pradesh, Manipur, Meghalaya, Mizoram, Tripura and the new states of Chattisgarh, Jharkhand, and Uttaranchal.

ANNEX 5

EXISTING GENERATING STATIONS OF CSUs, AS OF JUNE 2001

Region	Name of Station	Company	State/District	Installed Capacity (MW)	Plant Type	Energy Generated (Million of kWh in 1998/1999)
Northern Region	Singrauli	NTPC	UP / Sonebadhra	5 × 200 2 × 500	Coal	15,797.8
	Rihand	NTPC	UP / Sonebadhra	2 × 500	Coal	6,817.7
	Dadri	NTPC	UP / Gautam Budh Nagar	4 × 210	Coal	6,727.5
	Unchahar	NTPC	UP / Rai Bareilly	2 × 210	Coal	3,023.1
	ANTA	NTPC	Rajasthan / Baran	3 × 88 1 × 149	Gas	2,931.1
	Auraiya	NTPC	UP / Auraiya	4 × 110 2 × 106	Gas	4,146.2
	Dadri	NTPC	UP / Gautam Budh Nagar	4 × 131 2 × 146.5	Gas	5,099.2
	Baira Siul	NHPC	HP / Chamba	3 × 60	Hydro	750
	Chamera	NHPC	HP / Chamba	3 × 180	Hydro	2,367
	Tanakpur	NHPC	UP / Udhamsingh Nagar	3 × 31.4	Hydro	469
	Salal	NHPC	J&K / Udhampur	6 × 115	Hydro	3,222
	Uri	NHPC	J&K / Baramulla	4 × 120	Hydro	2,575
	RAPS	NPCIL	Rajasthan / Rawatbhata	1 × 100 1 × 200 2 × 220	Nuclear / PHWR	1,865
	NAPS	NPCIL	UP / Narora	2 × 220	Nuclear / PHWR	2,808
North Eastern Region	Assam	NEEPCO	Assam / Bokuloni	6 × 33.5 3 × 30	Natural Gas	743.3
	Agartala	NEEPCO	Tripura / Ramchandranagar	4 × 21	Natural Gas	197.2
	Kopili. Khandong Power House	NEEPCO	Assam / Hills	2 × 25 2 × 50	Hydro	556.5
	KHEP Stage – 1 Extension Koplili power house	NEEPCO	Assam / Hills	2 × 25 2 × 50	Hydro	438.7
	Loktak hydroelectric project	NHPC	Manipur / Bishanpur & Churachandpur	3 × 35	Hydro	532
Eastern Region	Farakka	NTPC	West Bengal / Mushirabad	3 × 200 2 × 500	Coal	5,475.6
	Kahalgaon	NTPC	Bihar / Bhagalpur	4 × 210	Coal	3,988.7
	Talcher	NTPC	Orissa / Angul	2 × 500	Coal	4,592.5
	Talcher (old)	NTPC	Orissa / Angul	4 × 60 2 × 110	Coal	2,248.5
Western Region	TAPS	NPCIL	Maharashtra / Tarapur	2 × 160	Nuclear / BWR	2,294
	KAPS	NPCIL	Gujarat / Kakrapar	2 × 220	Nuclear / PHWR	2,894
	Korba	NTPC	MP / Jamnipali	3 × 200 3 × 500	Coal	16,046.6
	Vindhychal	NTPC	MP / Sidhi	6 × 210	Coal	9,934.2
	Kawas	NTPC	Gujarat / Surat	4 × 106 2 × 110.5	Gas	4,411.9
	Jhanor-Gandar	NTPC	Gujarat / Bharuch	3 × 131 1 × 255	Gas	2,162.2
Southern Region	Thermal Power Station – I	NLC	Tamil Nadu / Neyveli Cuddalore	6 × 50 3 × 100	Lignite	3,772.2
	Thermal Power Station – II	NLC	Tamil Nadu / Neyveli Cuddalore	7 × 210	Lignite	9,568.1
	Ramagundam	NTPC	AP / Karimnagar	3 × 200 3 × 500	Coal	15,859.2
	Kayamkulam	NTPC	Kerala / Allepey	2 × 115	Gas	177.8
	Kaiga	NPCIL	Karnataka / Kaiga	2 × 220	Nuclear / PHWR	1,853
	MAPS	NPCIL	Tamil Nadu / Kalpakkam	2 × 170	Nuclear / PHWR	2,188

Sources: NPCIL: http://www.npcil.org.   *Others: http://www.cercind.org*

ANNEX 6

INSTALLED GENERATING CAPACITY AND GROSS ENERGY GENERATION OF SEBS, AS OF 31 MARCH 1998

Region	SEB	Installed Capacity (Mw)	Plant Type	Energy Generated (GWh)
Northern Region	Haryana	883.9	Hydro	3,919.9
		892.5	Steam	3,342.1
		3.9	Diesel & Wind	0
		0	Gas	0
		1,780.3	*Total*	*7,262*
	Himachal Pradesh	299.2	Hydro	1,285.4
		0	Steam	0
		0.1	Diesel & Wind	0
		0	Gas	0
		299.3	*Total*	*1,285.4*
	Jammu & Kashmir	190.2	Hydro	735
		0	Steam	0
		8.4	Diesel & Wind	0
		175	Gas	63
		374.1	*Total*	*798*
	Punjab	1,798.9	Hydro	8,666.6
		1,920	Steam	8,232.1
		0	Diesel & Wind	0
		0	Gas	0
		3,718.9	*Total*	*16,898.7*
	Rajasthan	971.1	Hydro	3,977.2
		975	Steam	5,935.1
		0	Diesel & Wind	0
		38.5	Gas	16.3
		1,984.6	*Total*	*9,928.7*
	Uttar Pradesh	1,504.8	Hydro	5,014
		4,664	Steam	17,813.3
		0	Diesel & Wind	0
		0	Gas	0
		6,168.8	*Total*	*22,827.3*
		14,326	**Sub-Total**	**59,000**
Western Region	Gujarat	487	Hydro	738
		3,804	Steam	21,212
		18.5	Diesel & Wind	0
		198	Gas	1,103
		4,507.5	*Total*	*23,053*
	Madhya Pradesh	847.9	Hydro	2,253.2
		3,017.5	Steam	15,345.7
		0.6	Diesel & Wind	0
		0	Gas	0
		3,866	*Total*	*17,598.9*
	Maharashtra	1,314.2	Hydro	3,381.7
		6,005	Steam	31,213.5
		4.6	Diesel & Wind	0
		912	Gas	4,875.1
		8,235.8	*Total*	*39,470.2*
		16,609.3	**Sub-Total**	**80,122.1**

Region	SEB	Installed Capacity (Mw)	Plant Type	Energy Generated (GWh)
Southern Region	Andhra Pradesh	2,519	Hydro	6,131.9
		2,952.5	Steam	15,102.9
		3	Diesel & Wind	0.2
		0	Gas	0
		5,474.6	*Total*	*21,234.9*
	Karnataka	206.2	Hydro	617
		0	Steam	0
		127.9	Diesel & Wind	543
		0	Gas	0
		334.1	*Total*	*1,160*
	Kerala	1,676.5	Hydro	6,626.1
		0	Steam	0
		82	Diesel & Wind	0
		0	Gas	0
		1,758.5	*Total*	*6,626.1*
	Tamil Nadu	1,955.7	Hydro	4,714.5
		2,970	Steam	17,219.7
		19.4	Diesel & Wind	22.9
		130	Gas	17.9
		5,075	*Total*	*21,975*
		12,642.3	**Sub-Total**	**50,996**
Eastern Region	Bihar	150	Hydro	280.1
		1,393.5	Steam	1,997.1
		0	Diesel & Wind	0
		0	Gas	0
		1,543.5	*Total*	*2,277.2*
	West Bengal	127.2	Hydro	132.2
		1020	Steam	3,061.6
		12.6	Diesel & Wind	0
		100	Gas	14.7
		1,259.3	*Total*	*3208.5*
		2,802.7	**Sub-Total**	**8,155.5**
North Eastern Region	Assam	2	Hydro	0
		330	Steam	693.9
		20.7	Diesel & Wind	0
		244.5	Gas	728.8
		597.2	*Total*	*1,422.7*
	Meghalaya	186.7	Hydro	542.6
		0	Steam	0
		2	Diesel & Wind	0
		0	Gas	0
		188.8	*Total*	*542.6*
		785	**Sub-Total**	**1,965.3**
All India (all SEBs)		**47,166.3**	**TOTAL**	**200,238.9**

Source: Central Electricity Authority, 1997/98. Public Electricity Supply, All India Statistics, General Review, New Delhi.

ANNEX 7

FULLY-COMMISSIONED PRIVATE POWER PROJECTS, AS OF 31 JANUARY 2001

	Name / Location	Promoters	Net Capacity (MW)	Technology	Fuel	Project Cost (Rs. billion)	Situation
1	Paguthan / Gujarat	Torrent Group and Powergen	655	CCGT	Natural gas / Naphtha	23	Fully commissioned (with TEC)
2	Hazira / Gujarat	Essar Power	515	CCGT	Natural gas / Naphtha	17	Fully commissioned (with TEC)
3	Baroda / Gujarat	GIPCL	167	CCGT	Naphtha / HSD	4	Fully commissioned (with TEC)
4	Surat Lignite / Gujarat	GIPCL	2 × 125 = 250	TPS	n.a.	12	Fully commissioned (with TEC)
5	Dabhol / Maharashtra	Dabhol power Co.	2,184 (740 + 1444)	CCGT	Natural gas / Naphtha	28	Fully commissioned (with TEC)
6	Jegurupadu / Andhra Pradesh	GVK Industries	216	CCGT	Natural gas / Naphtha	8	Fully commissioned (with TEC)
7	Godavari / Andhra Pradesh	Spectrum Power Generation	208	CCGT	Natural gas / Naphtha	7	Fully commissioned (with TEC)
8	Basin Bridge / Tamil Nadu	GMR Vasavi Power Corp.	4 × 50 = 200	DGPP	LSHS	8	Fully commissioned (with TEC)
9	Toranagallu / Karnataka	Jindal Group	2 × 130 = 260	TPS	n.a.	11	Fully commissioned (with TEC)
10	Kondapally / Andhra Pradesh	Kondapally Power Corp.	350	CCGT	Naphtha	10	Fully commissioned (with TEC)
11	Guntur Branch Canal-I / Andhra Pradesh	n.a.	3.75	HEP	n.a.	n.a.	Fully commissioned (without TEC)
12	Shivpur / Karnataka	Bhoruka Power Company	18	HEP	n.a.	n.a.	Fully commissioned (without TEC)
13	Maniyar / Kerala	Carborandum Universal	12	HEP	n.a.	n.a.	Fully commissioned (without TEC)
14	Reliance Salgaocar Project / Goa	n.a.	48	n.a.	n.a.	n.a.	Fully commissioned (without TEC)
15	Adamtilla / Assam	DLF Power Co.	9	n.a.	n.a.	n.a.	Fully commissioned (without TEC)
16	Bansakandi / Assam	DLF Power Co.	15.5	n.a.	n.a.	n.a.	Fully commissioned (without TEC)
17	Gurgaon / Haryana	Magnum Power	25	CCGT	n.a.	n.a.	Fully commissioned (without TEC)
18	Tawa / MP	Hindustan Electro Graphite	13.5	HEP	n.a.	n.a.	Fully commissioned (without TEC)
19	Bellary Power Project / Karnataka	n.a.	27.8	n.a.	FO/LSHS	n.a.	Fully commissioned (without TEC)
20	Eloor / Kerala	BSES Kerala Power	173	CCGT	n.a.	n.a.	Fully commissioned (without TEC)

	Name / Location	Promoters	Net Capacity (MW)	Technology	Fuel	Project Cost (Rs. billion)	Situation
21	New Southern Gen. Station / Calcutta	CESC	135	n.a.	n.a.	n.a.	Licensees
22	Trombay / Maharashtra	BSES	180	TPS	n.a.	n.a.	Licensees
23	Dahanu / Maharashtra	BSES	500	TPS	n.a.	n.a.	Licensees
24	Bhira / Maharashtra	Tata Electric Company	150	n.a.	n.a.	n.a.	Licensees
25	Budge-Budge / W. Bengal	CESC	500	TPS	n.a.	n.a.	Licensees
	Total		**5371.55**				

Sources: http://www.gipcl.com
http://powermin.nic.in/nrg71.htm
http://cea.nic.in/opt4_tec.htm

ANNEX 8

WEB SITES ON INDIA

Central Governement

- Ministry of Power: powermin.nic.in
- Central Electricity Authority: www.cea.nic.in
- Central Electricity Regulatory Commission: www.cercind.org
- Power Grid Corporation of India: www.powergridindia.com
- Power Finance Corporation (PFC): www.pcfindia.com
- Power Trading Corporation (PTC): www.ptcindia.com
- Ministry of Coal: coal.nic.in
- Ministry of Petroleum and Natural Gas: petroleum.nic.in
- Ministry of Non Conventional Energy Sources: mnes.nic.in
- Indian Renewable Energy Development Agency: www.ireda.nic.in
- Department of Atomic Energy: www.dae.gov.in
- Ministry of Mines: www.nic.in/mines
- Ministry of Water Resources: wrmin.nic.in
- Ministry of Finance: finmin.nic.in
- Planning Commission: planningcommission.nic.in

State Governments

Some State Electricity Boards:

- Gujarat State Electricity Board: www.gseb.com
- Himachal Pradesh State Electricity Board: www.hpseb.com
- Maharashtra State Electricity Board: www.msebindia.com
- Tamil Nadu Electricity Board: www.tneb.org

Some State Electricity Regulatory Commissions:

- Andhra Pradesh Electricity Regulatory Commission: ercap.org

- Gujarat Electricity Regulatory Commission: www.gercin.org

- Haryana Electricity Regulatory Commission: herc.nic.in

- Karnataka Electricity Regulatory Commission: www.kar.nic.in/kerc

- Maharashtra Electricity Regulatory Commission: mercindia.com

- Orissa Electricity Regulatory Commission: www.orierc.org

- Uttar PRadesh Electricity Regulatory Commission: www.uperc.org

Some State governments:

- Government of Gujarat: www.gujaratindia.com

- Government of Orissa: www.orissagov.com

- Government of Uttaranchal: www.utaranchalassembly.org

- Government of Rajasthan, Department of Energy: www.rajenergy.com

Companies / Associations

- National Hydroelectric Power Corporation: www.nhpcindia.com

- National Thermal Power Corporation: www.ntpc.co.in

- Neyveli Lignite Corporation: www.nlcindia.co.in

- North Eastern Electric Power Corporation: www.neepco.com

- Nuclear Power Corporation of India: www.npcil.org

- Coal India: www.coalindia.nic.in

- Rural Electrification Corporation: rec.nic.in

- Council of Power Utilities: www.indiapower.org

- Confederation of Indian Industry: www.ciionline.org

- Federation of Indian Chambers of Commerce and Industry: www.ficci.com

- The Associated Chambers of Commerce and Industry of India: www.assocham.org

- Gas Authority of India: gail.nic.in

General Information

- Centre for Monitoring Indian Economy: www.cmie.com

- www.indiainfoline.com

- www.indiapoweronline.com

- Indian Electricity Portal: www.indianelectricity.com, www.indiaelectricitymarket.com

- Tata Energy Research Institute: www.teriin.org

- World Energy Council (India Member Committee): www.indiaworldenergy.org

- Asian Development Bank: www.adb.org

- World Bank: www.worldbank.org

ANNEX 9

ABBREVIATIONS AND ACRONYMS

ABT	Availability-based Tariff
ADB	Asian Development Bank
AEC	Ahmedabad Electricity Company
AES	Alternative Energy System Inc.
APGENCO	Andhra Pradesh Generation Company
APL	Adaptable Programme Loan
APSEB	Andhra Pradesh State Electricity Board
APTRANSCO	Andhra Pradesh Transmission Company
ASCI	Administration Staff College of India
bbl	barrel
BOT	Build Operate Transfer
BSEB	Bihar State Electricity Board
BSES	Bombay Suburban Electric Supply
bt	billion tonnes
BWR	Boiling Water Reactor
CCGT	Combined-cycle Gas Turbine
CCPP	Combined-cycle Power Plant
CDM	Clean Development Mechanism
CEA	Central Electric Authority
CERC	Central Electricity Regulatory Commission
CESC	Calcutta Electricity Supply Company
CESCO	Central Electricity Supply Company of Orissa
CIDA	Canadian International Development Agency
CMIE	Centre for Monitoring Indian Economy
CO_2	Carbon Dioxide
Crore	Ten million
CSU	Central Sector Utility
CTU	Central Transmission Utility
DFID	Department for International Development
DPC	Dabhol Power Company
DPL	Durgapur Projects Ltd
DVC	Damodhar Valley Corporation
EPC	Engineering, Procurement and Construction
ERC Act	Electricity Regulatory Commissions Act
ESMAP	Energy Sector Management Assistance Programme
FAC	Fuel Adjustment Charge
FSA	Fuel Supply Agreement
FTA	Fuel Transportation Agreement

GDP	Gross Domestic Product
GENCO	Generation Company
GIPCL	Gujarat Industries Power Company
GoUP	Government of Uttar Pradesh
GRIDCO	Grid Corporation of Orissa
GW	Gigawatt
HEP	Hydro Electric Power
HSD	High-Speed Diesel
HT	High Tension
HVDC	High Voltage Direct Current
IEA	International Energy Agency
IEGC	India Electricity Grid Code
IFC	International Finance Corporation
IGCC	Integrated Gasification Combined Cycle
IPP	Independent Power Producer
J&K	Jammu and Kashmir
KAPS	Kakrapar Atomic Power Station
KESCO	Kanpur Electricity Supply Company
km	kilometre
kV	kilovolt
kWh	kilowatt-hour
LNG	Liquefied Natural Gas
LPG	Liquefied Petroleum Gas
LSHS	Low Sulphur Heavy Stock
Ltd	Limited
MAPS	Madras Atomic Power Station
mbd	million barrels per day
MERC	Maharashtra Electricity Regulatory Commission
MESB	Maharashtra State Electricity Board
MoA	Memorandum of Agreement
MoP	Ministry of Power
MoU	Memorandum of Understanding
MP	Madhya Pradesh
MSEB	Maharashtra State Electricity Board
Mtoe	Million Tonnes of Oil Equivalent
MW	Megawatt
NAP	Naphta
NAPS	Narora Atomic Power Station
NCAER	National Council of Applied Economic Research
NEEPCO	North-Eastern Power Corporation
NHPC	National Hydro Power Corporation
NLC	Neyveli Lignite Corporation
NPC	Nuclear Power Corporation
NREB	Northern Region Electricity Board
NRLDC	Northern Region Load Dispatch Centre
NTPC	National Thermal Power Corporation
O&M	Operation and Maintenance

OECD	Organisation for Economic Co-operation and Development
OERC	Orissa Electricity Regulatory Commission
OHPC	Orissa Hydel Power Corporation
OPGC	Orissa Power Generation Corporation
OSEB	Orissa State Electricity Board
PFC	Power Finance Corporation
PCIL	Power Grid Corporation of India
PHWR	Pressurised Heavy Water Reactor
PLF	Plant Load Factor
POWERGRID	Power Grid Corporation of India
PPA	Power Purchase Agreement
PTC	Power Trading Corporation
R&M	Renovation and Modernisation
RAPS	Rajasthan Atomic Power Station
REB	Rural Electricity Board
REC	Rural Electrification Corporation
RECI	Regional Electric Co-operation and Integration
RLDC	Regional Load Dispatch Centre
Rs	Rupees
RSEB	Rajasthan State Electricity Board
SEB	State Electricity Board
SERC	State Electricity Regulatory Commission
SLDC	State Load Dispatch Centre
STN	Shutdown or standby
STU	State Transmission Utility
T&D	Transmission and Distribution
TAPS	Tarapur Atomic Power Station
TEC	Tata Electric Power Company
TEC	Techno-economic Clearance
TPES	Total Primary Energy Supply
TPS	Thermal Power Station
TRANSCO	Transmission Company
TWh	Terrawatt-hour
UP	Uttar Pradesh
UPERC	Uttar Pradesh Electricity Regulatory Commission
UPPCL	Uttar Pradesh Power Corporation
UPSEB	Uttar Pradesh State Electricity Board
USD	US Dollars
UT	Union Territory
WB	World Bank
WBPDC	West Bengal Power Development Corporation
WBSEB	West Bengal State Electricity Board
WEC	World Energy Council
WEO	World Energy Outlook

ORDER FORM

IEA BOOKS

Fax: +33 (0)1 40 57 65 59
E-mail: books@iea.org
www.iea.org/books

INTERNATIONAL ENERGY AGENCY

9, rue de la Fédération
F-75739 Paris Cedex 15

I would like to order the following publications

PUBLICATIONS	ISBN	QTY	PRICE		TOTAL
☐ **Electricity in India - *Providing Power for the Millions***	92-64-19724-9		$125	€137	
☐ World Energy Outlook - *2001 Insights*	92-64-19658-7		$150	€165	
☐ Energy Policies of IEA Countries - 2001 Review (Compendium)	92-64-19659-5		$120	€132	
☐ International Emission Trading - *From Concept to Reality*	92-64-19516-5		$100	€110	
☐ Oil Supply Security - *The Emergency Response Potential of IEA Countries in 2000*	92-64-18575-5		$100	€110	
☐ Competition in Electricity Markets	92-64-18559-3		$75	€82	
☐ Energy Labels and Standards	92-64-17691-8		$100	€110	
☐ Energy Balances of Non-OECD Countries – 2001 Edition	92-64-08744-3		$110	€121	
			TOTAL		

DELIVERY DETAILS

Name _____ Organisation _____

Address _____

_____

Country _____ Postcode _____

Telephone _____ E-mail _____

PAYMENT DETAILS

☐ I enclose a cheque payable to IEA Publications for the sum of $_____ or €_____

☐ Please debit my credit card (tick choice).   ☐ Mastercard   ☐ VISA   ☐ American Express

Card no: ⌊_⌊_⌊_⌊_⌊_⌊_⌊_⌊_⌊_⌊_⌊_⌊_⌊_⌊_⌊_⌊_⌊_⌋

Expiry date: ⌊_⌊_⌊_⌊_⌋   Signature:

OECD PARIS CENTRE

Tel: +33 (0)1 45 24 81 67
Fax: +33 (0)1 49 10 42 76
E-mail: distribution@oecd.org

OECD BONN CENTRE

Tel: +49 (228) 959 12 15
Fax: +49 (228) 959 12 18
E-mail: bonn.contact@oecd.org

OECD MEXICO CENTRE

Tel: +52 (5) 280 12 09
Fax: +52 (5) 280 04 80
E-mail: mexico.contact@oecd.org

You can also send your order to your nearest OECD sales point or through the OECD online services:
www.oecd.org/ bookshop

OECD TOKYO CENTRE

Tel: +81 (3) 3586 2016
Fax: +81 (3) 3584 7929
E-mail: center@oecdtokyo.org

OECD WASHINGTON CENTER

Tel: +1 (202) 785-6323
Toll-free number for orders:
+1 (800) 456-6323
Fax: +1 (202) 785-0350
E-mail: washington.contact@oecd.org

International Energy Agency, 9 rue de la Fédération, 75739 Paris CEDEX 15
PRINTED IN FRANCE BY CHIRAT
(61 02 05 1 P) ISBN 92-64-19724-9 - 2002

Cover pictures : GettyImages / Chris Mellor, Photodisc.